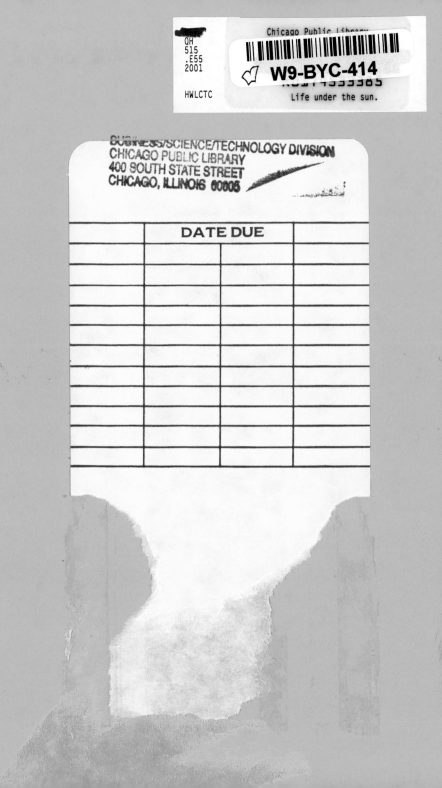

DATE DUE

Life Under the Sun

Life Under the Sun

Peter A. Ensminger

Yale University Press   New Haven and London

Designed by Sonia Shannon
Set in Cochin type by
The Composing Room of Michigan, Inc.
Printed in the United States of America by
R. R. Donnelley & Sons, Harrisonburg, Virginia.

Library of Congress Cataloging-in-Publication Data

Ensminger, Peter A., 1957–
Life under the sun / Peter A. Ensminger.
p.   cm.
Includes bibliographical references.
ISBN 0-300-08804-3 (cloth : alk. paper)
1. Photography.   2. Photoreceptors.   3. Light—
Physiological effect.   4. Vision.   I. Title.

QH515 .E55   2001
571.4′55 — dc21   00-043663

A catalogue record for this book is available from
the British Library.

The paper in this book meets the guidelines for
permanence and durability of the Committee on
Production Guidelines for Book Longevity of the
Council on Library Resources.

10 9 8 7 6 5 4 3 2 1

Contents

Preface

I wrote *Life Under the Sun* to share with people who have an enthusiasm for science in general and biology in particular my fascination with animals, plants, fungi, and microbes that respond to light. I hope that this book will appeal to professional biologists as well, for biology is such a diverse field that few professionals know all about the light-induced responses in the various organisms discussed in this book. A specialist is often just a layperson when outside his or her own discipline.

Peter Medawar has argued that the scientific paper is an exercise in deception because it misleads readers about the true thought processes that underlie scientific discovery. According to Medawar, scientists lead us to believe that their research is conducted in the same logical and sequential manner in which it is presented in their papers. True, nonscientists might be deceived. But no professional scientist is fooled, because we all practice this so-called deception regularly in our own papers.

In any case, this utterly undeceptive book is a collection not of scientific papers but of informal essays. In *Essais* (1580), the collection generally credited with introducing the form, Michel de Montaigne wrote, "I am myself the matter of my book." Although there is more of me in the essays of *Life Under the Sun* than would be appropriate in formal scientific papers, the organisms themselves are the matter of my book.

The first three essays delve into vision in humans and other animals. The next three consider the effects of light on human health. The topics then diverge to describe light effects in plants, fungi, and other organisms. There is no need for readers to progress sequentially through the essays. In fact, I anticipate that after taking in the Introduction, readers will migrate to the essays they find most personally interesting. One person may turn to

> ### Some Basic Photochemistry
>
> Any molecule that absorbs light or radiation is a potential chromophore — light-absorbing component — for photochemical reactions. Photochemistry begins when the absorption of radiation excites a molecule to a higher energy level. Sometimes this excitation energy is dissipated as heat or as light of a longer wavelength (fluorescence or phosphorescence); in this case, the energy has been "wasted," because it did not cause any permanent chemical changes. Under certain conditions, the excitation energy can cause chemical transformations of the molecule; in this case, the energy has performed "photochemical work." Some examples of photochemical work discussed in this book are decomposition (photolysis; see Chapters 4 and 8) and structural rearrangement (photoisomerization; see Chapters 1, 2, 7, and 14).

Chapter 5, "A SAD Tale," about the effects of light on Seasonal Affective Disorder; another may begin with Chapter 12, "Turning on a Butterfly," about special photoreceptors that are located on the genitalia of butterflies; and another may jump to Chapter 15, "Too Much of a Good Thing," about methods that plants use to cope with excessive sunlight.

Some of these essays have appeared elsewhere. Chapter 8, "Light and Beer," an account of the deleterious effect of light on beer, is adapted from an article published in *Zymurgy*, a magazine for home brewers (of which I am one). Chapter 7, "A Novel Method of Weed Control," which describes the use of nighttime plowing to control weeds, and Chapter 4, "A Burning Issue," about the harmful effects of ultraviolet radiation, were adapted from articles that originally appeared in *Biology Digest*, a monthly periodical for biology students. I am grateful to the publishers of these essays for permission to reprint the revised versions here.

Why essays? In *Night Life*, a collection of essays about wildlife at nighttime, Diana Kappel-Smith has compared writers to carpenters, some of whom are expert at making furniture, others at making buildings. Like her, I am more a maker of furniture than of buildings, and consequently my book, like hers, is a collection of essays. This probably has something to do with my beginning as a biologist at the laboratory bench, composing brief, narrowly focused articles. Although I have written in many other forms since my conversion to a biologist at the computer keyboard, I still feel most comfortable composing small pieces about specific topics. I hope that readers will appreciate this collection as a whole, much as we can appreciate the arts-and-crafts ambience in a room full of Stickley furniture.

Anyone whose appetite has been whetted by an essay can consider it a stepping stone to pursue the topic further by looking at the Suggested Reading list at the close of each essay. In addition, I have included a Glossary for readers who may be stymied by a technical term, and I have compiled extensive endnotes for readers who want to pursue the scientific literature. I have also prepared a Web site with supplementary information and lists of related sites. My site is accessible from the Yale University Press site for *Life Under the Sun* (www.yale.edu/yup/books/088043.htm) and will be regularly updated. I encourage you all to contact me through this site.

Acknowledgments

I must thank many individuals for their help in preparing this book. In addition to my wife, Lisa, who provided encouragement throughout, and our dog Sandy (now deceased), who lay at my side as I wrote, I feel lucky to have had constructive feedback from the foremost experts in the various fields covered. In particular, I thank Jeff Palmer (Indiana University) for discussing the phylogeny depicted in figure 1; Robert B. Barlow (State University of New York Upstate Medical Center, Syracuse) for discussing his research on horseshoe crabs and *Rimicaris* shrimp and for his comments on Chapter 1; Steve Chamberlain (Syracuse University) for providing videotapes of black smokers and *Rimicaris* shrimp, discussing *Rimicaris* shrimp with me, and commenting on Chapter 2; Thomas W. Cronin (University of Maryland, Baltimore County) for discussing his research on mantis shrimps and commenting on Chapter 3; Jack Werner (University of Colorado, Boulder) for discussing his work on impressionism and the vision of Claude Monet; Richard Setlow (Brookhaven National Laboratory) for his comments on Chapter 4; Betsy Sutherland (Brookhaven National Laboratory) for discussing photoreactivation in humans; Norman E. Rosenthal (National Institute of Mental Health, NIH, Bethesda) for commenting on Chapter 5; Martin J. Warren (University College, London) for commenting on Chapter 6; Karl M. Hartmann (Erlangen University, Germany) for help in preparing figure 2 and table 1 and for discussing his research on seed germination and commenting on Chapter 7; Denis De Keukeleire (University of Ghent, Belgium) for discussing his research on hop bitter acids and commenting on Chapter 8; Morten Meilgaard (Stroh Brewery), John Paul Maye (Pfizer), and David Hysert (John I. Haas) for discussing photostable hop compounds; Rudy Held (Kalsec) for sending literature on photostable hop compounds; David Dennison (Dartmouth College) for discussing his experience as a student in Max Delbrück's lab and for commenting on Chapter 9; Paul R. Fisher (LaTrobe University, Melbourne, Australia) and Donat P. Häder (Erlangen University, Germany) for discussing their research on *Dictyostelium;* Paul R. Fisher for commenting on Chapter 10; Pill Soon Song (University of Nebraska, Lincoln) for commenting on Chapter 11; Kentaro Arikawa (Yokohama City University, Japan) for discussing his research on butterfly genitalia and for commenting on Chapter 12; Woody Hastings (Harvard University) for commenting on Chapter 13; Jeff Stuart (Syracuse University) for discussing his research on *Halobacterium* and commenting on Chapter 14; and William W. Adams III (University of Colorado, Boulder) for commenting on Chapter 15. Lastly, I thank my editors at Yale University Press: Dan Heaton, for his meticulous attention to detail, and Jean Thomson Black, for gently ushering me into the world of book publishing.

Introduction

Of Physiology from top to toe I sing
— *Walt Whitman*

The earth, which is 93 million miles from the sun, receives a small fraction of the light that the sun radiates into space, but that small fraction serves as the energy source that supports life on our planet. People have long appreciated the great significance of the sun. Many ancient cultures in Africa, Asia, and the Americas put it at the center of their religions. Many other cultures, both ancient and modern, have recognized that the sun provided them with the light and warmth that sustained them and their crops, and have used the sun as a cultural or religious motif.[1]

Modern science has provided us with a more complete understanding of why the sun is so important (see sidebar, "Some Basic Photochemistry," p. viii). Plants use chlorophyll and other pigments to absorb the energy in sunlight, and the biochemical reactions of photosynthesis transform solar energy into biochemical energy. Plants then use this biochemical energy for growth, while giving off oxygen as a waste product. Animals eat plants, or they eat other animals that eat plants, and they breathe in the oxygen expelled by the plants.

But sunlight provides more than the *energy* for photosynthesis, and thereby the food we need to eat and the oxygen we need to breathe. Sunlight also provides *information* to plants, animals, fungi, and microbes about their environments. Thus life on earth — which has evolved under the influence of sunlight for billions of years — has become specialized to sense the quantity, direction, polarization, wavelength, and periodicity of light. Sunlight allows animals to see, and it controls movement and morphogenesis in plants, fungi, and microbes. It tells organisms the time of day and the time of year by resetting their circadian and annual rhythms. But sunlight can also be injurious: the ultraviolet radiation in sunlight harms many animals by suppressing their immune systems and it can damage the DNA of nearly all organisms.

These essays do not catalogue all the different biological responses to light; nor do they provide a comprehensive coverage of photobiology, the branch of biology that deals with the interactions of light and living organisms. Many of the books listed under Suggested Reading provide a more complete coverage of photobiology and more complete descriptions of the myriad biological effects of light.

Instead, this book presents intriguing pieces of the overall picture by describing specific responses to light in diverse organisms, providing a cross-section of photobiology. These responses include *vision*, the perception of the overall appearance of objects that is mediated by the eyes and brains of animals; *photosynthesis*, the conversion of light energy into biochemical energy that is used for food and occurs in plants and some lower life forms; *phototaxis*, the movement of an organism according to the direction of light that occurs in many single-celled organisms; *phototropism*, the orientation of a sessile organism or one of its parts according to the direction of light; and *photosensitization*, a reaction to light that can lead to red and swollen tissues in humans and other animals. This collection of essays describes these and other responses to light in a diversity of organisms whose evolutionary relationships are depicted in figure 1. By sampling these very different responses in these very different organisms, the reader gets a glimpse of the larger picture, for nature often reveals itself in its individual creations.

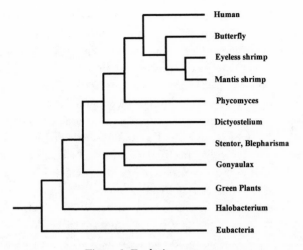

Figure 1. Evolutionary tree.
The evolutionary branching pattern of organisms
featured in *Life Under the Sun*, based on data presented
by Joel Dacks and Andrew J. Roger (J. Dacks and A. J.
Roger, 1999, The first sexual lineage and the relevance of
facultative sex, *Journal of Molecular Evolution* 48, 779–83).
This tree groups together related organisms and shows, for
example, that the mantis shrimp and eyeless shrimp are
more closely related to each other than either is to the
yellow swallowtail butterfly. Branch lengths do not reflect
evolutionary distance.

Levels of Light on Earth

Light can be measured in different ways and with different types
of instruments. Photobiologists use many technical terms to
characterize environmental light levels because there are so
many different types of biological responses to light.[2] Thus researchers studying vision, photosynthesis, phototaxis, phototropism, and photosensitization must use different types of light
measurements. Unfortunately, some of the terminology used to

describe environmental light levels has changed over time, so different researchers may use the same term for slightly different types of measurements.[3] Adding to the confusion, some photobiology research papers do not specify the aspect of a light field that has been measured, nor the methods used for measurement.

The indivisible unit of light or electromagnetic radiation is the photon. One simple way to express environmental light levels is *photon flux*, which is defined as the number of photons that strike a flat surface of a specified area in a given unit of time. A similar measurement is *spherical photon flux*, which is defined as the number of photons that strike a sphere of a specified cross-section per unit of time. Spherical photon flux, which is often used by plant biologists, takes into account the three-dimensional structure of organisms that can sense light from all directions. Under the special circumstance in which all light comes from a single direction — which is often the case in laboratory experiments — the photon flux is equal to the spherical photon flux. Under natural conditions, however, the spherical photon flux is higher than the photon flux, and the relation between the two values varies according to the time of day.[4] In a somewhat Procrustean manner, table 1 classifies many of the diverse responses described in the essays of this book according to spherical photon flux.

Throughout the course of a day, the range of the spherical photon flux typically encountered on earth spans seven or eight orders of magnitude.[5] At the high end of this scale — noon on a clear day — most of the light on earth obviously comes from the sun. But what about the small amount of ambient light present at nighttime, long after sunset? Aside from artificial sources, this residual light comes from the moon, the stars, and air glow, a natural phenomenon caused by chemical reactions driven by electron currents in the upper atmosphere.[6]

The enormous range of light levels that are encountered during the day is similar in magnitude to the difference between the height of a small child (about 1 meter) and the diameter of the earth (about 12,800 kilometers). Because we are — in the words of vision scientist and Nobel laureate Haldan K. Hartline — "children of the sun," it should not be surprising that our own eyes can function over this entire range.[7] In fact, our eyes function over an even broader range, and so do the light-sensing systems of many other organisms (table 1).

Wavelengths of Light on Earth

We are children of the sun in more ways than one. Not only do our eyes and the photosensory systems of many other organisms function over the range of light levels present on earth, but we and other organisms are also most sensitive to the range of wavelengths that are brightest on earth, between 400 and 700 nm (see figure 2).[8] As Newton showed with a simple prism, the white light from the sun is composed of many colors and extends from violet to blue, green, yellow, orange, and then red. Very little ultraviolet radiation below 300 nm reaches the earth because the sun's emission falls off very sharply below 300 nm and because stratospheric ozone absorbs much of the radiation in this region of the spectrum. Very little infrared radiation above 1000 nm reaches the earth because the emission from the sun declines at long wavelengths and because water in the earth's atmosphere absorbs strongly above 1000 nm.[9]

Although most photobiological responses are elicited by light between 400 and 700 nm, there are significant differences in the pigments used to absorb light and in the particular wavelengths that these pigments most effectively absorb (see figure 2). Thus certain butterflies have photoreceptors on their genitalia that absorb maximally at about 380 nm (ultraviolet radiation); during daytime, human vision is most sensitive to yellow-green light of about 555 nm; and photosynthesis in higher plants is most effectively stimulated by red light of about 675 nm. Three different photoreceptive molecules are responsible for these different

Table 1

Levels of light on earth and biological responses

Column 1 lists different environmental light levels;[1] column 2, spherical photon flux (number of visible photons [400 to 700 nm] intercepted by a sphere of one meter cross-section per second) associated with these light levels;[1] column 3, biological responses associated with these light levels; column 4, photoreceptive pigments for the different biological responses; column 5, chapters that discuss the biological responses; and column 6, notes and references for the biological responses.

Environmental light level	Spherical photon flux (photons m^{-2}s^{-1})	Biological response	Photoreceptive pigment	Chapter	Notes and references
Clear midday	~10^{21}	Photosynthesis saturation, sun-loving plants (continuous exposure to sunlight)	Chlorophyll	15	Light level is species-dependent[2]
Cloudy midday	~10^{20}	Photosynthesis saturation, shade-loving plants (continuous exposure to sunlight)	Chlorophyll	15	Light level is species-dependent[2]
Clear sunset	~10^{19}	Lengthening of *Gonyaulax* circadian period (single four-hour red-light pulse)	Chlorophyll (?)	13	Threshold for bioluminescence rhythm[3]
Early twilight, clear	~10^{18}	Seed germination (LFR) (one-second red-light pulse)	Phytochrome	7	Threshold for low-fluence response[4]
Late twilight, clear	~10^{17}	Negative phototaxis in *Dictyostelium* amoebas (continuous white-light exposure)	Porphyrin (?)	10	Higher light levels: negative phototaxis (AX-2 strain)[5] Lower light levels: positive phototaxis (AX-2 strain)[5]

~10^{16}	Photophobic response in *Stentor coeruleus* (continuous red-light exposure)	Stentorin	11	Threshold level for light avoidance in "stella" strain[6]
~10^{15}	Human color vision	Iodopsin	1, 2	Threshold corresponding to ~10^{-2} milli-Lamberts[7]
~10^{14}	Seed germination (VLFR) (one-second red-light pulse)	Phytochrome	7	Threshold for very low–fluence response[4]
~10^{12}	Positive phototaxis in *Dictyostelium* slugs (continuous white-light exposure)	Porphyrin (?)	10	Threshold level (NC4 strain)[8] Threshold for AX2 strain is ~100-fold lower[9]
~10^{10}	Human vision (one-millisecond green-light pulse)	Rhodopsin	1	Threshold corresponding to ~$10^{-6.5}$ milli-Lamberts[7,10]
~10^{9}	Phototropism in *Phycomyces* (continuous blue-light exposure)	Flavin	9	Threshold level[11]
~10^{7}	Adaptation acceleration in *Phycomyces* (35-minute red-light pulse)	Flavin (?)	9	Threshold level[12]

1. Based on measurements in Erlangen, Germany given in K. M. Hartmann et al., 1998, Photocontrol of germination by

(*continued*)

Table 1 Continued

moon and starlight, *Zeitschrift Pflanzen Krankheit Pflanzen Schutz*, Sonderheft 16, 119–27; K. M. Hartmann, 1995, Harrowing at night is half weeded, *International Symposium on Weed and Crop Resistance to Herbicides*. Córdoba, Spain. Simultaneous measurements of spherical illuminance, a measure of the light level sensed by the human eye, range from about 10^5 lux during a clear midday to about 10^{-3} lux during a cloudy midnight at new moon.

2. F. B. Salisbury and C. W. Ross, 1985, *Plant Physiology*, Belmont, Calif., Wadsworth.

3. T. Roenneberg and J. W. Hastings, 1991, Are the effects of light on phase and period of the Gonyaulax clock mediated by different pathways? *Photochemistry and Photobiology* 53, 525–33.

4. K. M. Hartmann and W. Nezadal, 1990, Photocontrol of weeds without herbicides, *Naturwissenschaften* 77, 158–63.

5. D.-P. Häder and B. Vollersten, 1991, Phototactic orientation in *Dictyostelium discoideum* amoebae, *Acta Protozoologica* 30, 19–24.

6. C. B. Hong et al., Light-adaptation in the photophobic response by *Stentor coeruleus*, *Archives of Microbiology* 147, 117–20.

7. P. Buser and M. Imbert, 1995, *Vision*, Cambridge, MIT Press.

8. K. L. Poff and D.-P. Häder, 1984, An action spectrum for phototaxis by pseudoplasmodia of *Dictyostelium discoideum*, *Photochemistry and Photobiology* 39, 433–36.

9. D.-P. Häder and A. Hansel, 1991, Responses of *Dictyostelium discoideum* to multiple environmental stimuli, *Botanica Acta* 104, 200–205

10. S. Hecht et al., 1942, Energy, quanta, and vision, *Journal of General Physiology* 25, 819–40.

11. K. Bergman et al., 1969, *Phycomyces*, *Bacteriological Review* 33, 99–150; P. A. Ensminger et al., 1990, Action spectra for photogravitropism in *Phycomyces* wild type and three behavioral mutants (L150, L152, and L154), *Photochemistry and Photobiology* 51, 681–87.

12. P. Galland et al., 1989a, Subliminal light control of dark adaptation kinetics in *Phycomyces* phototropism, *Photochemistry and Photobiology* 49, 485–91; X.-Y. Chen et al., 1993, Action spectrum for subliminal light control of adaptation in *Phycomyces* phototropism, *Photochemistry and Photobiology* 58, 425–31.

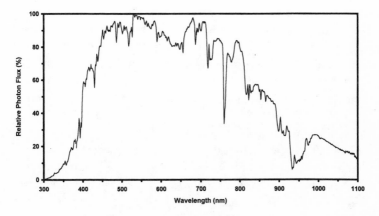

Figure 2. Daylight spectrum.

The spectrum of sunlight on earth, from 300 to 1100 nm, scanned in 1 nm steps with 2 nm bandwidth and recorded on a clear summer day (August 7, 1991) by Karl Hartmann and Wolfgang Kaufmann at 12:00 Central European Time on the flat roof of the Biologikum at Erlangen University, Germany (49°35′N, 11°02′E; 300 m above sea level). The wavelength maxima of biological responses discussed in *Life Under the Sun* are:

~260 nm — DNA damage, *in vitro* (Chapter 4)

~280 nm — Photo-degradation of isoalpha acids in beer (Chapter 8)

~310 nm — Sunburn (Chapter 4)

~380 nm — Genitalic photoreception in butterflies (Chapter 12)

~410 nm — Phototaxis in *Dictyostelium* slugs and amoebas (Chapter 10)

~450 nm — Phototropism in plants and *Phycomyces* (Chapter 9)

~475 nm — *Gonyaulax* bioluminescence (Chapter 13)

~505 nm — Human vision at low light levels (Chapters 1, 2, 3)

~555 nm — Human vision at high light levels (Chapters 1, 2, 3)

~568 nm — Photosynthesis in *Halobacterium* (Chapter 14)

~610 nm — Photophobic response & negative phototaxis in *Stentor* (Chapter 11)

~668 nm — Formation of the Pfr form of phytochrome (Chapter 7)

~675 nm — Photosynthesis in land plants (Chapter 15)

~730 nm — Formation of the Pr form of phytochrome (Chapter 7) responses. The chemical structures of all the photoreceptive molecules discussed in this book appear in the appendix. Even readers with little training in chemistry can look at these figures and appreciate the basic similarity of these molecules, for they all have one or more ringlike structures and alternating single and double bonds between carbon atoms, which chemists call "conjugated double bonds."

Our own eyes and the eyes of other animals do not simply measure light's wavelength; they judge colors according to wavelength combinations. Plants can also sense wavelength combinations with phytochrome, a photosensory pigment that behaves like photochromic eyeglasses, which change color under different types of light. Certain fungi and microbes may also be able to sense wavelength combinations. Plants, fungi, and microbes lack nervous systems, so none could be considered to have color vision. But the diverse pigments and the exquisite specializations that have evolved for absorbing different wavelengths of light evince the importance of light's wavelength.

Light is so important to us that Diane Ackerman has called our eyes "the great monopolists of our senses."[10] The same might be said of the photosensory and energy-transducing systems of the many organisms discussed in this book, for they truly seem to dominate the lives of these organisms.

Suggested Reading

Björn, L. O. (1976) *Light and Life*. New York, Crane, Russak.

Buser, P., and M. Imbert (1992) *Vision*. Cambridge, MIT Press.

Fein, A., and E. Z. Szuts (1982) *Photoreceptors: Their Role in Vision*. New York, Cambridge University Press.

Kohen, E., et al. (1995) *Photobiology*. New York, Academic Press.

Mast, S. O. (1911) *Light and the Behavior of Organisms*. New York, Wiley.

Smith, K. C., ed. (1989) *The Science of Photobiology*. New York, Plenum.

Wolken, J. J. (1995) *Light Detectors, Photoreceptors, and Imaging Systems in Nature*. New York, Oxford University Press.

Wyszecki, G., and W. S. Stiles (1982) *Color Science*. New York, Wiley.

1 *Vision at the Threshold*

Obviously the amount of energy required
to stimulate any eye must be large enough
to supply at least one quantum to the
photosensitive material. No eye need be
so sensitive as this. But it is a tribute to
the excellence of natural selection that
our own eye comes so remarkably close to
the lowest limit.

— *Selig Hecht, Simon Shlaer,*
and Maurice Pirenne

Vision is surely the most important of our senses. A single glance instantly gives us information about our surroundings that is much more sophisticated than that from our other senses. Our eyes perform many amazing feats. They focus on objects at different distances, adapt to ambient light levels that vary more than a billionfold in brightness, and discriminate among many different colors, from the violet to the red.[1] One of the most remarkable features of our eyes is that they can function at extraordinarily low light levels.

How much light do we need to see? Experiments performed under certain controlled conditions have shown that we can perceive a one-millisecond flash of blue-green light that contains about 100 photons.[2] Other experiments have shown that we can perceive a steady point source of blue-green light that has a flux of about 800 photons per second.[3] Because the photosensitive cells of the retina absorb only about 10 percent of the blue-green light that actually enters the eye, it follows that the threshold for vision occurs when the retina absorbs about 10 photons from a one-millisecond flash or about 80 photons per second from a continuous point source of light.[4]

The retinas of certain animals that are renowned for their night vision absorb a higher percentage of the light that enters their eyes because they have larger pupils, which allow more light to reach the retina, and reflective tapetums at the back of their eyes, which reflect light that was not absorbed during its first pass through the photoreceptor cells.[5] Thus cats and owls have a threshold that is about five times lower than our own, even though their retinas have about the same photosensitivity.

What limits our sense of vision at low light levels? This simple question is not so simply answered. Indeed, vision scientists have seriously investigated this question since the late nineteenth century.[6]

We know that at low light levels, such as on a starry night (see table 1, p. 6), vision is due mostly to light absorption by the visual pigment rhodopsin in the rod cells of the retina. Each rhodopsin molecule is composed of a retinal molecule (a vitamin A analog) that is attached to a protein called opsin (see figure A1, Appendix). Each of our rod cells contains about 100 million molecules of rhodopsin, and the retina of each eye has about 100 million rod cells. Thus each of our eyes has about 10^{16} (ten thousand trillion) molecules of rhodopsin. Having so much rhodopsin is a definite advantage for vision at low light levels, for it increases the probability that a photon that enters the eye will be captured. We also know that the key event in vision is a rapid light-induced activation (photoisomerization) of the rhodopsin molecule.[7] The efficiency of this process is also a definite advantage for vision at low light levels.

To really understand what limits our sense of vision at low light levels, we need to look at a few key experiments that have been performed in the past fifty years.

Detection of a Single Photon

One of the best known experiments in vision science was performed at Columbia University in the 1940s by Selig Hecht, Simon Shlaer, and Maurice Pirenne.[8] These researchers gave one-millisecond flashes of light of variable fluence rate ("intensity") to one another and a few volunteers under highly controlled conditions. The light flashes were administered following thirty minutes of dark adaptation and were so narrowly focused that they fell upon only about 500 of the more than 100 million rod cells in each observer's retina. Hecht, Shlaer, and Pirenne then recorded how frequently the different observers perceived the flashes of light.

They used a relatively simple mathematical method, known as Poisson statistics, to analyze the data. Their results showed that a person can perceive a light flash if 500 nearby rod cells of the retina absorb about six photons within one millisecond. If six photons are absorbed by 500 rod cells, then the probability that any single rod cell has absorbed two photons is very small ($P = 6/500 \times 6/500$, or about one in seven thousand). Thus Hecht, Shlaer, and Pirenne concluded that activation of about six nearby rod cells within one millisecond is sufficient for the perception of a light flash.

These experiments are renowned because they demonstrated very simply and elegantly — yet indirectly — that a single photon that is absorbed by a single molecule of rhodopsin can chemically excite a rod cell. Unfortunately, some students and biologists who are not well-versed in vision science occasionally misunderstand these results to mean that we can actually perceive a single photon. This is not at all correct. In fact, rhodopsin molecules in several nearby rod cells must absorb photons for the perception of a light flash. Moreover, the photons must be absorbed within a very small area of the retina and must be administered within a critical period of time. If a larger area of retina is stimulated, or the timing of the light flash is significantly greater than one millisecond, then more photons are necessary to trigger light perception.

A limitation of these experiments is that they are indirect. How can it possibly be proven that a single photon can excite a single rod cell?

The answer to this question came in the early 1960s from electrical recordings of the retina of the horseshoe crab *(Limulus polyphemus)*.[9] In particular, experiments performed on horseshoe crabs maintained in complete darkness showed that their retinas exhibit random electrical fluctuations, similar to the static on an untuned radio. Illumination increases the number of these fluctuations, a bit like turning up the volume on the same untuned radio. Analysis of the effect of the fluence rate of light on the electrical noise of the horseshoe crab retina showed that absorption of a single photon causes a single "bump" of electrical noise. This was an important confirmation of the conclusion of Hecht, Shlaer, and Pirenne.

Unfortunately, such experiments could not be performed with human retinas, because the rod cells of vertebrates are electrically coupled ("linked"). Denis Baylor and colleagues at Stanford University overcame this difficulty in 1979 by developing the "suction-electrode method," in which the outer segment of a single rod cell is sucked into the tip of a very small pipette. This elegant method allows measurement of the electrical current from isolated rod cells. The original experiments were performed with rod cells from the toad *Bufo marinus*, but subsequent experiments were performed with rod cells from primates.[10] The results showed that absorption of a single photon by a single rhodopsin molecule electrically activates a rod cell by blocking the entry of about one million positively charged ions into the cell. This further confirmed the conclusion of Hecht, Shlaer, and Pirenne that human rod cells respond to the absorption of a single photon.

A New Question Arises

Just as these important electrophysiological experiments answered one important question, they gave rise to another. What causes the electrical noise that occurs in rod cells held in complete darkness?

Early vision scientists realized that whatever the source of this noise, it was significant to the visual process.[11] Rod cells seem to have a basal level of electrical fluctuations, also known as "dark light," and illumination merely increases the number of these fluctuations.[12] The disadvantage of this electrical noise is that it places a lower limit on the amount of light that can be perceived. But there is also an advantage of this electrical noise. If excitation of a single rod cell, rather than several rod cells, led to perception of light, then we would constantly be confusing the electrical noise inherent to the rod cells with actual light signals.

Examination of the eyes of different animals provides a hint about the source of this electrical noise. Among different animal species, there is a strong correlation between the electrical noise level and the rhodopsin content of the photosensitive rod cells.[13] This suggests that rhodopsin itself is the source of the noise. Moreover, different species appear to have evolved different mechanisms for reducing the level of electrical noise in their photoreceptor cells. Humans and other primates have relatively small rod cells; small cells have less rhodopsin and are therefore less noisy.[14] Many cold-blooded animals, such as amphibians, have very large photoreceptor cells; they achieve low noise levels

by living at low temperatures, where the noise level is greatly reduced.[15] This gives cold-blooded animals better night vision than
humans and most other warm-blooded animals.

An early hypothesis was that the electrical noise in rod cells
is caused by "thermal activation" of rhodopsin, the visual pigment.[16] An explanation of this early hypothesis went something
like this: Rhodopsin and all other molecules have a certain
amount of thermal energy that causes them to bend and twist
about at random. Every now and then, the thermal energy of a
rhodopsin molecule is large enough to cause an activation of the
same sort that occurs following absorption of a photon. The
probability of thermal activation in any single rhodopsin molecule is extremely small. Each rod cell has 100 million rhodopsin
molecules, however, so the probability of thermal activation occurring anywhere in a single rod cell is much greater.

There is a problem with this thermal activation hypothesis.
Thermodynamic measurements showed that light activation of
rhodopsin required crossing a large energy barrier (activation
energy, 45 kcal per mole), whereas thermal activation of rhodopsin required crossing a small energy barrier (activation energy,
25 kcal per mole).[17] Thus the mechanism that causes electrical
fluctuations of rod cells kept in darkness is different from that
which causes electrical fluctuations following illumination.

Robert Barlow and colleagues from Syracuse University
and Rockefeller University have proposed a novel solution to this
problem.[18] Light-activation of a rod cell has long been known to
involve activation of rhodopsin, a molecule that consists of retinal
bound to an opsin protein by a "protonated Schiff-base linkage"
(see figure A2, Appendix). Based on theoretical calculations and
experimental results, Barlow and colleagues suggest that dark
activation of rhodopsin occurs as a two-step process:

1. The positive charge between the retinal and opsin
   is removed, creating a bond that lacks a positive

charge, known technically as a "deprotonated Schiff-base linkage."

2. This rare, deprotonated form of rhodopsin is thermally activated, thus causing electrical noise in darkness.

This controversial hypothesis predicts that increasing the proton concentration (acidity) inside photoreceptor cells should increase the amount of positively charged (protonated) rhodopsin, decrease the amount of uncharged (deprotonated) rhodopsin, and therefore reduce the electrical noise level of the cell. Thus the researchers tested their hypothesis by performing experiments on the horseshoe crab, the same organism that was previously used to demonstrate that absorption of a single photon causes a single "bump" of electrical noise.

The brain of the horseshoe crab normally generates a circadian rhythm that includes transmission of nerve signals to the photoreceptor cells, reducing their noise level at nighttime. Thus in their first test, Barlow and colleagues mimicked this brain signal by delivering electrical current shocks to the photoreceptor cells and noted a rapid increase in the proton concentration of the photoreceptor cells and a subsequent reduction in photoreceptor noise level. In their second test, they injected the eye of the horseshoe crab with a mildly acidic solution and noted a rapid and dramatic reduction in the electrical noise level. Both of these results are consistent with the researchers' novel hypothesis for the generation of photoreceptor noise.[19]

A more recent study has concluded that the same novel mechanism is the most likely source of noise in the photoreceptor cells in the eyes of all invertebrates and vertebrates, including those of humans.[20] Indeed, the rod cells of some vertebrates become more acidic with roughly the same time course as the decrease in overall electrical noise.[21] Thus stabilization of the positively charged form of rhodopsin during prolonged exposure to darkness appears to contribute to the high photosensitivity that humans and other vertebrates attain during dark adaptation.

A rod cell in the retina of our eyes can sense the absorption of a single photon, and absorption of photons by several rod cells allows us to perceive a flash of light. Because the photon is the indivisible unit of light energy, our eyes have achieved a nearly optimal solution for detection of low levels of light. In a review of the different sensory systems of many different organisms, William Bialek showed that many of these sensory systems also approach the limits imposed upon them by the laws of physics for detection of low stimulus levels.[22]

This degree of biological perfection may seem surprising, because some evolutionary biologists have maintained that natural selection does not allow organisms to achieve perfect solutions but allows only solutions that are compromises, solutions that result from many trials and errors over the course of many generations.[23]

As indicated by the classic paper of Hecht, Shlaer, and Pirenne, examination of the physical limits to the perception of light has not only taught us a great deal about the biochemistry and physiology of vision, it has also taught us something important about evolution and natural selection. Perhaps we can learn more about biology by examining how the laws of physics constrain the function of organisms as well as individual biological molecules.

Suggested Reading

Barlow, R. B. et al. (1993) On the molecular origin of photoreceptor noise. *Nature* 366, 64–66.

Bialek, W. (1987) Physical limits to sensation and perception. *Annual Review of Biophysics and Biophysical Chemistry* 16, 455–78.

Birge, R. R., and R. B. Barlow (1995) On the molecular origins of thermal noise in vertebrate and invertebrate photoreceptors. *Biophysical Chemistry* 55, 115–26.

Hecht, S., et al. (1942) Energy, quanta, and vision. *Journal of General Physiology* 25, 819–40.

2 The Five Percent Solution to Vision

*To suppose that the eye, with all its
inimitable contrivances for adjusting the
focus to different distances, for admitting
different amounts of light, and for the
correction of spherical and chromatic
aberration, could have been formed by
natural selection, seems, I freely confess,
absurd in the highest degree.*

— *Charles Darwin*

So begins a famous passage entitled "Organs of Extreme Perfection and Complication" in *The Origin of Species*.[1]

Half a century before Darwin, the theologian William Paley made an eloquent exposition of what has become known as the "argument from design" in his book *Natural Theology*.[2] Paley argued that just as a watch is evidence of a watchmaker, the eye and other functionally complex organs are evidence of God, a supreme creator. The Scottish philosopher David Hume, however, criticized the argument from design a full century before Darwin by arguing that intricately designed organs cannot provide positive evidence for a supreme creator.[3] Unfortunately, Hume did not offer an alternative explanation for the eye and other functional complexity in nature.[4]

In 1859 Charles Darwin gave an alternative explanation

that once and for all refuted Paley's argument from design.[5] Darwin demonstrated that the designlike features of the human eye and other complex organs in animals and plants are the consequence of evolution by natural selection. In the specific case of the human eye, Darwin realized that eyes do not always make very good fossils, so he argued that we cannot simply look at its lineal progenitors to understand how it evolved. Instead, he argued that the evolution of the eye by natural selection would be easier to accept if we could find animals with different types of eyes, from the very simple to the very complex, and if each type of eye could be shown to aid its possessor. By looking at the eyes of other animals, Darwin argued, we might better understand how our own eyes evolved.

Vision in Lower Life Forms

Indeed, there is a rich panoply of organisms from which to draw examples. The vision of humans and all other animals is based on light absorption by rhodopsin.[6] All rhodopsins use a vitamin A analog as the chromophore (light-absorbing component), and this is bound to a protein called opsin (figure A1, Appendix). The gene sequence and structure of opsins are remarkably similar among the many diverse organisms that have vision of one sort or another.[7]

Halobacterium salinarum, a halophilic (salt-loving) archaebacterium, is a primitive single-celled organism that has four different rhodopsin-like pigments (see Chapter 14). Two of these (bacteriorhodopsin and halorhodopsin) are used to drive photosynthesis, the conversion of light energy into chemical energy; the other two (sensory rhodopsin I and II) are sensory pigments that direct the cell's whiplike flagella to move it toward red and orange light, the same colors that are most effective in driving photosynthesis. The sequences of the *Halobacterium* rhodopsin genes suggest that they are not homologous with animal rhodopsins. But animal and *Halobacterium* bacteriorhodopsins have similar structures (seven helical traverses of the cell's membrane), and both use vitamin A analogs as chromophores, so the sequence dissimilarity may merely reflect the extensive evolutionary divergence of these proteins over the past billion years or so.[8]

Rhodopsin-based photoreceptors are also found in many species of green algae (Chlorophyta), a more modern group of organisms.[9] The green alga with the most intensively studied rhodopsin-mediated response is *Chlamydomonas reinhardtii*, a single-celled flagellated alga.[10] *Chlamydomonas*, like *Halobacterium*, uses rhodopsin to orient itself toward light so it can maximize photosynthesis. Unlike *Halobacterium*, though, *Chlamydomonas* uses chlorophylls, not rhodopsins, as photosynthetic pigments.

Invertebrates also use rhodopsin as a visual pigment.[11] It has been estimated, however, that photoreceptor cells have evolved independently at least forty separate times in invertebrates.[12] The vision of different invertebrates is based on many different optical design principles, and they have eyes like pinhole cameras, eyes with simple cameralike lenses, and at least five different types of compound eyes.[13]

One of the most primitive visual systems is found in *Nautilus*, a squidlike sea-dwelling mollusk that lives in a shell. Its eyes function like pinhole cameras and form images simply by shading.[14] Its eyes lack lenses, even though lenses would markedly brighten and sharpen the dim and blurry image that typically results. The eyes of *Nautilus* clearly demonstrate that even crudely built and partially functional eyes are better than no eyes at all.[15]

Five Percent of an Eye

A familiar query of those who argue that the eye is evidence for the hand of God and against natural selection goes something like: "What good is five percent of an eye?" In other words, with all the different parts of the eye so intricately complex and so interdependent, how could a partially formed eye, a presumed precursor of the human eye, be of any benefit whatsoever? The evolutionary biologist Richard Dawkins termed this the "argument from personal incredulity."[16] In effect, its proponent is saying, "I cannot possibly imagine how an incomplete and crudely formed eye could benefit an organism." Dawkins considers such an argument to be based on nothing more than its proponent's own ignorance of biological diversity.[17] In fact, if natural selection favored an increase in the amount of spatial information collected by a flat patch of photosensitive epidermis on an animal, then optical considerations lead to the estimate that well-developed eyes with lenses can evolve in several hundred thousand years, a mere twinkling of evolutionary time.[18]

In 1989 biologists described what might be called the "five percent eye" in a marine arthropod known as *Rimicaris exoculata* — the eyeless shrimp (see figure 3).[19] More recently, biologists have identified related species with similar features.[20] *Rimicaris* acquired its specific epithet — *exoculata* — because, unlike its Caridean shrimp relatives, it does not have eyes on stalks. *Rimicaris* lives in an unimaginably inhospitable environment, next to 350° C. (approximately 660° F.) hydrothermal vents that lie two miles beneath the surface of the Atlantic Ocean. A brief description of the hydrothermal vent environment will help illuminate the unique environment where *Rimicaris* shrimp live.

Hydrothermal vents, also called "black smokers" because they look like giant chimneys, were first discovered in 1977 in the Pacific Ocean near the Galapagos Islands.[21] These geological formations spew vast amounts of heat and minerals from the earth's interior into the ocean. Peter Rona and colleagues discovered hydrothermal vents in the Atlantic Ocean in 1985.[22] The Atlantic vents are found in the Mid-Atlantic Ridge, which runs north-south through the Atlantic Ocean, about 1,800 miles from the east coast of North America.[23] These vents form because of

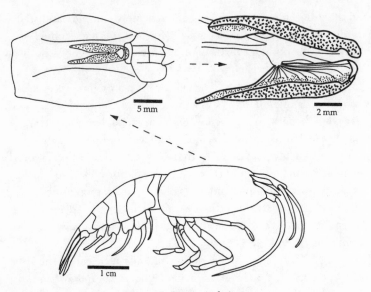

Figure 3. *Rimicaris* shrimp.
Bottom: side view of *Rimicaris exoculata,* showing the lack of eyestalks.
Top left: top view of the head region, with the spotted area showing the
location of the novel rhodopsin-containing organs ("eyes") that may
function in vision. Top right, dissected "eyes," showing the underlying
neural connections. Drawings courtesy of Steven C. Chamberlain,
Syracuse University.

the upwelling of molten lava between the earth's tectonic plates
as they slide past one another. On video, the vents actually appear to be emitting billows of black smoke, like giant furnaces.[24]

Tubeworms and bivalves dominate the vents of the Pacific.
These animals have symbiotic bacteria within their bodies that
oxidize sulfur, apparently providing their animal hosts with nutrition.[25] In contrast to the fauna of the Pacific geysers, *Rimicaris*
shrimp are the dominant animal species of Atlantic geysers.[26]
Videos show that many thousands of these shrimp swarm to-

gether near hydrothermal vent openings, feeding on the bacteria that thrive in these mineral-rich waters.[27]

Hydrothermal vents support a rich diversity of unique organisms. The photosynthetic plankton, bacteria, and plants that serve as primary producers in nearly all other ecosystems on earth cannot survive in the deep, dark ecosystems that surround hydrothermal vents. Instead, chemoautotrophic bacteria serve as the primary producers. These bacteria live on hydrogen sulfide, carbon dioxide, and other nutrients that are pumped up from the earth's interior through the vents. The vent bacteria oxidize hydrogen sulfide to sulfate and use the energy from this oxidation process to synthesize glycogen (a complex carbohydrate) from carbon dioxide.[28] This is one of the few food chains on earth that does not depend directly upon sunlight.

It seems amazing that a shrimp or any other animal could survive in the harsh environment of a hydrothermal vent. Steve Chamberlain and colleagues speculate that a shrimp species migrated from the ocean surface to the deep ocean in the past 5,000 to 10,000 years and that these immigrants eventually evolved into a new species, *Rimicaris exoculata*.[29] *Rimicaris* lacks the stalked eyes typical of its Caridean shrimp relatives, such as *Palaemonetes pugio* and *Palaemonetes vulgaris*, which have true eyes that are mounted on eyestalks.[30]

Recent work suggests that the eyeless shrimp is not quite eyeless. Researchers of the Woods Hole Oceanographic Institution, Syracuse University, and the University of Newcastle have found that *Rimicaris* has a pair of bright reflective surfaces on its back, between its gill chambers.[31] Dissection indicated that rudimentary eyelike organs lie within the body, beneath these reflective surfaces. These unusual "eyes" do not have image-forming optics, but they underlie a transparent region of the shell. Examination of the ultrastructure of these internal organs showed that they are composed of the same clusters of visual elements (ommatidia) that make up the compound eyes of invertebrates and closely resemble the eyes of other shrimp species. In fact, the "eyes" of *Rimicaris* shrimp have roughly the same number of visual elements as the stalked eyes of *Palaemonetes pugio* and *Palaemonetes vulgaris*.[32] The clusters of visual elements in *Rimicaris*, however, are much larger than those of *Palaemonetes*, making its "eyes" potentially capable of detecting much lower levels of light.[33]

Further study of *Rimicaris* has shown that each rudimentary "eye" has about 10,000 photoreceptor-like cells and an optic nerve that connects these photoreceptors to the brain.[34] There is also a highly reflective layer on the back of each cluster of photoreceptor cells. Such reflective layers are relatively common in nocturnal animals. Cats, for example, are well known for their "eyeshine," a phenomenon wherein a reflective layer at the back of each eye makes it appear to shine like a mirror when illuminated at nighttime.[35] Eyeshine allows light that was not absorbed by the photoreceptors during its passage into the eye to pass back through the photoreceptors again on its way out, so that it has a second chance of being absorbed by the photoreceptor pigments.

Perhaps most significant of all, Cindy Lee Van Dover and colleagues have shown that the rudimentary eyes of *Rimicaris* shrimp have high concentrations of rhodopsin, the same pigment that humans and all other animals use for vision. Each eye of *Rimicaris* has about 25 picomoles (1.5×10^{13} molecules) of rhodopsin, about the same as the eyes of the much-studied horseshoe crab, *Limulus polyphemus*.[36] The horseshoe crab is also a marine arthropod, but is not a true crab of the class Crustacea (see Chapter 1).

The presence of specialized cells that have such high concentrations of rhodopsin certainly suggests that the eyeless shrimp can see. This is difficult to prove, however, because scientists cannot easily perform the requisite behavioral tests. *Rimicaris* lives so deep beneath the ocean surface that scientists can observe it only through the window of a specially designed deep submergence vehicle near a thermal vent. Any scientist who attempted to exit the vehicle to perform behavioral experiments would be crushed by the enormous water pressure or burned by the tremendous heat of a black smoker. How can we tell whether *Rimicaris* shrimp can really see? And just what is there to be seen next to a hydrothermal vent that lies two miles beneath the ocean surface? At this depth, there is no sunlight and no bioluminescent creatures that can be seen by dark-adapted human observers.

Can the Eyeless Shrimp Really See?

Denis Pelli and Steve Chamberlain of Syracuse University hypothesized that *Rimicaris* uses its rudimentary eyes to sense thermal vents and that this ability helps it to feed on the bacteria that thrive in these nutrient-rich waters.[37] It is difficult for scientists to observe light-guided behavior in *Rimicaris,* because the artificial light necessary to observe from the deep submergence vehicle surely disrupts the shrimp's visual system and probably blinds it, unaccustomed as the shrimp is to such intense illumination.[38]

Because behavioral experiments are impracticable, Pelli and Chamberlain investigated this question from a theoretical and biophysical perspective.[39] In short, they asked whether any physical constraints existed that would prevent the eyes of *Rimicaris* shrimp from sensing a thermal vent.

First, they made the simplifying assumption that the heat that emanates from a thermal vent is similar to that emitted by a "black-body." According to a theory developed by the physicist Max Planck, a black-body is matter that emits radiation throughout the wavelength spectrum, extending from the infrared, through the visible, into the ultraviolet.[40] As the temperature of a black-body increases, the wavelength of maximal emission decreases. As a bar of iron is heated in a furnace, for example, its color changes from ruby red to bright red to orange to yellow; as an electric stove element heats up, its color changes from ruby red to bright red to orange.

Then, Pelli and Chamberlain calculated that a 350° C. thermal vent emits radiation maximally at 5889 nm, deep in the infrared region of the spectrum. The level of radiation emitted in the visible region is about 10^{10} to 10^{15} times lower. The rhodopsin in the eyes of *Rimicaris* shrimp absorbs maximally near 500 nm, and its absorbance tails off very sharply through the red and infrared regions.[41]

By multiplying the emission spectrum of the thermal vent and the absorption spectrum of rhodopsin, Pelli and Chamberlain calculated that *Rimicaris* sees a hydrothermal vent as emitting radiation with a maximum at 588 nm. Thus, even though the emission spectrum of the hydrothermal vent and the absorbance spectrum of rhodopsin differ substantially, their spectral overlap indicates that *Rimicaris* rhodopsin absorbs vent radiation in the visible region of the spectrum. Pelli and Chamberlain then wanted to determine if there was enough radiation from the hydrothermal vent to stimulate photoisomerization, the light-mediated structural change in rhodopsin that is the first biochemical event of vision (see figure A1, Appendix). By accounting for the concentration of rhodopsin in the eyes of *Rimicaris* and various other optical factors, they estimated that a 350° C. hydrothermal vent will cause about 1,800 rhodopsin photoisomerizations per second in the eyes of *Rimicaris* when the vent fills the shrimp's visual field (technically, when the angular subtense of the target is $\pi/2$). This is sufficient to support vision.

A question may naturally arise. If *Rimicaris* shrimp can see a thermal vent, then can we humans? Pelli and Chamberlain had the same question. Thus, while two miles beneath the ocean surface in their deep submergence vehicle, the researchers allowed their own eyes to become dark-adapted (to increase their photosensitivity), then looked at the thermal vent. They saw nothing.

Based on certain biophysical calculations, however, Pelli and Chamberlain determined that the radiation emitted by a hydrothermal vent was just below the visual threshold for humans.

Lest anyone be suspicious of such calculations, as I myself once was, a simple experiment can be done to demonstrate that our own eyes can sense the radiation emitted from a black-body. All I needed was a laboratory hot plate and a very dark basement to test the Pelli and Chamberlain conjecture.

First, I went into my basement at nighttime and set the temperature of a laboratory hot plate to about 400° C. (750° F.). I made the room as dark as possible, so dark that not a single speck of light could be seen. I sat in the room for about twenty minutes and allowed my eyes to fully adapt to the darkness. Finally, I moved close to the hot plate and looked at it. I saw a diffuse glow! The same theory used to predict that *Rimicaris* shrimp sees hydrothermal vents also predicts that humans should see a glow at about 590 nm (yellow-orange). This simple experiment supports that theory.

In relating the story of the Eyeless Shrimp, I am reminded of a story told by a witness to the first nuclear bomb test in south-central New Mexico. The witness told of a relative who was blind but nevertheless could sense the intense light that was given off by the explosion, which was about a thousand times brighter than the brightest summer day.[42] If the eyes of this blind person and the "five percent eyes" of *Rimicaris* shrimp can actually see — even crudely — then the evolution of the eye by natural selection does not seem absurd at all.

In the words of Darwin, "In living bodies, variation will cause slight alterations, generations will multiply them almost infinitely, and natural selection will pick out with almost unerring skill each improvement. Let this process go on for millions of years; and during each year on millions of individuals of many kinds; and may we not believe that a living optical instrument might thus be formed?"[43]

Suggested Reading

Darwin, C. (1859) *The Origin of Species*. London, Murray; rpt. 1958, New York, New American Library.

Dawkins, R. (1987) *The Blind Watchmaker.* New York, Norton.

Goldsmith, T. (1990) Optimization, constraint, and history in the evolution of eyes. *Quarterly Review of Biology* 65, 281–322.

Nilsson, D.-E., and S. Pelger (1994) A pessimistic estimate of the time required for an eye to evolve. *Proceedings of the Royal Society of London* B 256, 53–58.

Pelli, D. G., and S. C. Chamberlain (1989) The visibility of 350° C blackbody radiation by the shrimp *Rimicaris exoculata* and man. *Nature* 337, 460–61.

Rona, P. A. (1992) Deep-sea geysers. *National Geographic*, October, 105–9.

Van Dover, C. L. (1996) *The Octopus's Garden: Hydrothermal Vents and Other Mysteries of the Deep Sea*. Reading, Mass., Addison-Wesley.

Van Dover, C. L., et al. (1989) A novel eye in "eyeless" shrimp from hydrothermal vents of the Mid-Atlantic Ridge. *Nature* 337, 458–60.

3 A More Delightful Vision

Everything that lives strives for color.
—*Johann Wolfgang von Goethe*

The colors we see depend on the wavelength sensitivities of the visual receptors within our eyes as well as the wavelengths of light that enter our eyes. In color vision light excites different classes of photoreceptor cells, containing different visual pigments, and the brain compares their differential light absorption.[1] Thus in bright light humans see a colorful world because the cone cells in our retinas have three visual pigments, with maximal sensitivities in the blue (~425 nm), green (~530 nm), and red (~560 nm) regions of the spectrum, and the differential responses of these cells enables color vision.[2] But our own colorful world pales in comparison with the world of the mantis shrimps. These crustaceans must be considered the champions of color vision, because their eyes have more than ten different classes of visual pigments.[3]

The importance of our visual pigments in determining the perception of color is perhaps best illustrated by "color-blindness." This genetic disorder, which occurs in about one of twelve males and one of one hundred females, is caused by a defective visual pigment gene or the loss of a gene that codes for the red or green visual pigment. Although people with this disorder perceive the world very differently, they are not truly color-blind, for they still have two bright-light visual pigments and use these for dichromatic color vision, a more rudimentary type of color vision. True color-blindness (achromatopsia) occurs in people who lack both the red and green visual pigments; this condition is extremely rare, occurring in fewer than one person in thirty thousand.[4]

Molecular biologists have shown that there are small differences in the gene sequences of the visual pigments in different people and that this variation is widespread.[5] Such genetic variation occurs even among those of us with "normal" color vision. Because the gene sequence of a visual pigment determines its wavelength sensitivity, it follows that the eyes of different people differ in their wavelength sensitivities. In fact, based on the measured variation of visual pigments, geneticists have estimated there is a probability of about 50 percent that any two randomly chosen Caucasian men have different perceptions when looking at the same color.[6] The situation among females is more complex and has not yet been elucidated.

There are presumably additional variations among people in the neural circuitry that processes visual information. Variations also occur as individuals age, as illustrated by the work of the impressionist painter Claude Monet. Monet's perception of color changed as he aged because he developed cataracts, and this had a profound influence on his painting style (see sidebar, "Claude Monet's Vision"). The important point is that different people live in different perceptual worlds. In the words of Alexander Pope,

> *The diff'rence is as great between*
> *The Optics seeing, as the objects seen.*
> *All Manners take a tincture from our own,*
> *Or come discolour'd thro' our Passions shown.*

As might be expected, the differences between how humans and animals perceive colors are even greater than the differences among humans.

- Dogs and cats, like most mammals, use two visual pigments for dichromatic color vision.[7]

Vision in the Stomatopod Superfamilies

- Bathysquillidoidea (two families) — group of deep-water (bathypelagic) species that are not well known. Their eyes lack midbands and they apparently cannot sense light color or polarization.
- Erythrosquilloidea (one family with one species) — its eyes lack midbands.
- Squilloidea (two families) — group of species that burrow in soft substrates. Their eyes have midbands with two rows of ommatidia.
- Lysiosquilloidea (five families) — group that includes very large species, most of which burrow in coral reefs and "spear" their prey. Their eyes have midbands with six rows of ommatidia.
- Gonodactyloidea (eight families) — group of coral reef–dwelling species known for "smashing" its prey. Their eyes have midbands with six rows of ommatidia.**

recent analysis of the evolutionary relationships of stomatopods has placed the 350 different species into five taxonomic groups called superfamilies: the Erythrosquilloidea, Bathysquilloidea, Squilloidea, Lysiosquilloidea, and Gonodactyloidea (see sidebar, "Vision in the Stomatopod Superfamilies").[12]

Species in the Bathysquilloidea are little studied because they live deep in the ocean (*bathos* is Greek for "deep") and are rarely seen. These species have relatively simple eyes, with no color vision.[13] Obviously, there is little need for a sophisticated color vision system in their dark habitat. Indeed, most other animals that live in such dark environments also lack color vision.[14]

Most species in the Squilloidea inhabit turbid water and are

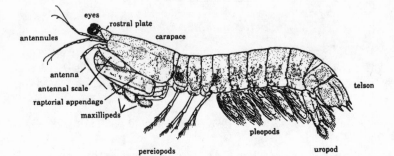

Figure 4. Mantis shrimp.
Top: side view of *Pseudosquilla ciliata* (subfamily, Gonodactyloidea). The
stalked eyes are located between the rostral plate, a beaklike extension of
the shell, and the antennules. Bottom: head-on view of *Odontodactyllus
scyllarus* (subfamily, Gonodactyloidea). Each eye has a dorsal and ventral
hemisphere and a six-rowed midband. The eye consists of thousands of
tiny optical units (ommatidia) that are packed together like the cells of a
beehive. The three dark areas on each eye (pseudopupils) allow trinocular
vision in each eye. Artwork courtesy of Roy Caldwell, University of
California, Berkeley.

nocturnal, and their eyes are only slightly more complex than those of the Bathysquilloidea. Like species of the Gonodactyl-oidea and Lysiosquilloidea, the Squilloids have compound eyes that are divided into three sections: a horizontal midband, a hemisphere above the midband, and a hemisphere below the midband (see figure 4). Species in the Squilloidea are among the largest of the stomatopods.

The Gonodactyloids and Lysiosquilloids have the most sophisticated eyes, and these species will be the focus of the rest of this essay. Most of these species live in the clear shallow water of brightly colored coral reefs and are active during the day.[15] They typically hide in coral reef burrows, sitting and waiting for their prey (worms, snails, fish, and other crustaceans), with only their stalked eyes protruding from the burrow entrance. They are known for aggressively attacking prey by smashing or spearing them with an arm. A strike is typically completed in three to four milliseconds and has an impact velocity of about ten meters per second, among the fastest known body movements of all animals.[16] These mantis shrimps are so violent that they even fight one another for territory or mates. In fact, when put into an aquarium, certain Gonodactyloid species are known to kill everything else inside and even break the glass walls, making them less than desirable as aquarium pets.[17]

Anyone who has snorkeled in a coral reef has seen its incredibly vibrant colors. Yet humans see these colors with only three visual pigments. A coral reef must be even more vibrant and colorful to a mantis shrimp, which has more than ten classes of visual pigments. It is possible that over time an evolutionary arms race has pitted the eyes of mantis shrimps against the unique colors that its prey use for camouflage.[18] Thus, it is possible that as its prey evolved ever more elaborate color schemes for camouflage, the eyes of mantis shrimps evolved ever more elaborate visual systems for detection of its camouflaged prey.

In addition to living in the colorful coral reef environment, mantis shrimps are themselves brightly colored, and many species have characteristic multicolored spots on their tail fins. Behavioral studies have shown that mantis shrimps use elaborate communication routines, which seem to require recognition of body colors with their uniquely sophisticated vision system.[19]

The human eye works like a camera, using a large light refractive lens to focus light onto a large photosensitive retina at the back of the eyeball. In contrast, the eyes of mantis shrimps and many other crustaceans are classified as apposition compound eyes. Apposition compound eyes are composed of many ommatidia, each with its own cornea and crystalline cone. Light rays must be aimed directly into an individual ommatidium to be sensed by the underlying visual pigments; the responses of thousands of these ommatidia are integrated to form an image. In the other major type of eye found in arthropods, superposition compound eyes, a cluster of nearby ommatidia direct their light rays onto the same patch of visual pigment.[20] Superposition eyes perform better at low light levels, but the apposition eyes of mantis shrimps have better resolution than many types of superposition eyes.

A truly remarkable feature of the compound eyes of mantis shrimps is that a single location in space can be seen by ommatidia in three different parts of the same eye — the upper hemisphere, midband, and lower hemisphere. Thus each eye of the mantis shrimp has depth perception and trinocular vision.[21] In contrast, humans require both of our eyes for depth perception and binocular vision. Aside from their roles in trinocular vision, the upper and lower hemispheres of the mantis shrimp compound eye are rather unremarkable. The ommatidia in these regions are used primarily for recognition of forms, not color vision.[22] Ommatidia in the upper and lower hemispheres contain two types of rhodopsin, one type that is sensitive to short wavelengths of light (such as blue), and another type that is sensitive to long wavelengths of light (such as red).[23] This is similar to the compound eyes of many other crustaceans.[24]

Sensation of Colors

The midband, which consists of six rows of ommatidia in the Gonodactyloids and Lysiosquillids, is the most remarkable region of the mantis shrimp's eyes.[25] The ommatidia of midband rows 1–4 are particularly important for color vision. Each of these has eight classes of visual pigments that vary in maximal sensitivity from about 400 nm (deep violet) to about 550 nm (green). Because the receptor cells that contain these pigments have special wavelength filters, the wavelength range of the different receptor cells is even greater, from about 400 nm to about 650 nm (ruby red). All these visual pigments use a vitamin A analog (11-cis retinal) as the light-absorbing compound that is attached to a protein called opsin (see figure A1, Appendix), so their wavelength sensitivities presumably differ because they are coded by different opsin genes.[26]

In addition to having a diversity of visual pigments, the ommatidia in midband rows 1–4 are segregated into three tiers. The ommatidia of rows 2 and 3 have two ranks of color filters that separate the three tiers, so that the color of light changes as it passes through these ommatidia. These color filters are mostly carotenoid compounds whose colors are purple, blue, red, orange, or yellow (depending on the species) and function in tuning and sharpening the wavelength range of light that reaches the visual pigments beneath.[27]

The color filters of the midband region, like sunglasses, markedly reduce the level of light that reaches the visual pigments, so mantis shrimps require bright light to use their sophisticated color vision systems. This may explain why they are so successful in the well-lit waters of coral reefs. The few Gonodactyloids and Lysiosquilloids that do live in deep or murky water generally lack certain types of color filters and some of the other adaptations for color vision.[28]

Detection of Light Polarization

Humans can detect light polarization, but our sensitivity is subtle. In fact, some vision scientists consider humans essentially blind to light polarization.[††] The polarized light that many people can see is evident in the presence and orientation of "Haidinger's brushes," which are visible when we look up at a white cloud through a polarizing filter or a lens of Polaroid sunglasses.[‡‡] Haidinger's brushes appear as a pair of fuzzy brushlike images that are yellow or brown on the sides and light blue on the top and bottom. Novices can learn to recognize the brushes by rotating a polarizing filter while looking up at a white cloud and noticing that the "brushes" also rotate. With practice, some people can even see Haidinger's brushes without a polarizing filter, by merely looking at the northern sky on a clear day. Our ability to detect light polarization is apparently due to the alignment of nerve fibers in the macula lutea (a small yellow area on the retina) which are dichroically oriented and absorb some light before it reaches the photoreceptor cells beneath.[§§]

Sensation of Light Polarization

While the ommatidia in midband rows 1–4 are used for color vision, those in rows 5 and 6 are used for detection of light polarization.[29] Our own eyes are poor at sensing light polarization (see sidebar, "Detection of Light Polarization") but very good at sensing its color and brightness; however, sensation of light polarization is not at all unusual in the animal kingdom. It is found in some insects, crustaceans, fish, birds, and particularly in cephalopods, a class of mollusks that includes squids, octopuses, and cuttlefish.[30] In analogy to color vision, a key requirement for polarization sensitivity is the excitation of two or more classes of visual pigments that have different alignments in the eye (technically, different axes of maximal excitation).[31] In the mantis shrimp, the visual pigments of the different tiers of optical elements in rows 5 and 6 have different alignments. Comparison of the neural inputs from these two rows allows the mantis shrimp to sense polarized light.[32]

The subtlety of polarization vision in humans makes it difficult for us to appreciate that polarized light signals are widespread throughout nature, including the shallow tropical waters inhabited by mantis shrimps. In general, polarization sensitivity enhances contrast vision, and this is particularly important for the mantis shrimp, because underwater objects generally have lower contrast.[33] The mantis shrimp's polarization sensitivity allows it to better see many of the fish upon which it preys, for the silvery sides of many fish species strongly polarize light.

Clearly, the most sophisticated region of the mantis shrimp's eye is the midband. The midband, however, views only a small area (about 5–10 degrees) of the visual field at any given instant.[34] Ommatidia in the upper and lower hemispheres, which lack the specializations for color and polarization vision, can sense only forms and motion. Thus it might seem that the mantis shrimp senses the color and polarization of light in only a very narrow range of its visual field.

But this is not quite correct. In fact, the eyes of the mantis shrimp are mounted on mobile stalks and are constantly moving about, apparently independent of one another.[35] These eye movements are controlled by eight individual eyecup muscles that have been divided into six functional groups.[36] Slow scanning movements of the eyes allow the mantis shrimp to "paint" color and polarization information onto the forms that are seen by ommatidia in the upper and lower hemispheres of its eyes.[37] In addition, mantis shrimp eyes, like those of humans, have saccadic movements.[38] These rapid eye movements facilitate rapid redirection of a gaze without blurring the visual image, just as readers of this essay are doing right now, moving from one line to the next. A third type of eye movement in the mantis shrimp is tracking, which is used in following moving objects, such as prey.[39] Tracking involves large, rapid eye movements, in which the two eyes appear to move independently, often up to 70 degrees, at an instant.

The eye of the mantis shrimp, with its trinocular vision, its multitude of visual pigments, its polarization sensitivity, and its intricate movements, is truly among the most specialized and most sophisticated eyes in the animal kingdom. How can humans possibly know what the mantis shrimp sees and what kind of world it lives in?

Alas, because of widespread variation in human visual pigment genes, we do not even share the same color perception with one another, so we can only imagine what the mantis shrimp sees. The sophisticated eyes of the mantis shrimp should teach us that visual perception is only relative and that our own view of the world is not necessarily the best view.

Suggested Reading

Cronin, T. W. et al. (1994) The unique visual system of the mantis shrimp. *American Scientist* 82, 356–65.

Jacobs, G. H. (1993) The distribution and nature of color vision among the mammals. *Biological Reviews* 68, 413–71.

The Lurker's Guide to Stomatopods (www.blueboard.com/mantis/). Includes many outstanding color photographs.

Nilsson, D.-E. (1989) From cornea to retinal image in invertebrate eyes. *Trends in Neurosciences* 13, 55–64.

Sacks, O. (1997) *Island of the Colorblind.* New York, Knopf.

Werner, J. S. (1998) Aging through the eyes of Monet. Pp. 3–41 in *Color Vision* (W. Backhaus et al., eds.). Berlin, Walter de Gruyter.

4 A Burning Issue

Mad dogs and Englishmen
go out in the midday sun.
—*Noël Coward*

Fair-skinned people who have had sunburns are personally familiar with at least one of the harmful effects of ultraviolet radiation. But all the media play given to the ozone hole, tanning, skin cancer, sunscreens, and related topics has created a lot of confusion and misunderstanding. Is the ozone layer really thinning, or is this just a hoax propagated by environmentalists? Is tanning under artificial sunlamps safer than tanning outdoors? What are the chances of developing skin cancer? Can sunscreens really protect against sunburns and skin cancer?

Ozone

The chemically heterogeneous ozone layer contains about 90 percent of the atmosphere's ozone (O_3) and lies in the lower stratosphere, about fifteen to fifty kilometers above the earth's surface.[1] This ozone is constantly being created and destroyed by chemical reactions that are driven by ultraviolet radiation from the sun. The air in the stratosphere is so rarefied that all its ozone would be about three millimeters thick if it were at the same pressure as air at the earth's surface.[2]

In 1881 William Hartley first showed that the sharp cutoff in terrestrial solar radiation in the ultraviolet region of the spec-

trum (~290 nm; see figure 2, p. 9) is due to absorption by ozone.[3]
We now recognize the biological significance of Hartley's observation. Stratospheric ozone absorbs ultraviolet radiation from
the sun that would otherwise damage the DNA of all organisms.
Ultraviolet radiation can also damage other biological molecules.
In fact, life on earth would be very different without the protective ozone layer.

Based on the different biological effects of the different
wavelengths of ultraviolet radiation, biologists commonly divide
this spectral region into three bands: ultraviolet-C, from 200 to
280 nm; ultraviolet-B, from 280 to 320 nm; and ultraviolet-A,
from 320 to 400 nm (see sidebar, The ABCs of the Ultraviolet).[4]
As Hartley showed, ultraviolet-C, though harmful, is mostly absorbed by the earth's atmosphere and is essentially undetectable
on earth. Most ultraviolet-B is screened out by ozone in the
stratosphere, but the small amount that reaches the earth can
have significant harmful effects. Ultraviolet-A is not as effectively screened out by molecules in the atmosphere, and its level
is about 1,000 times as high as ultraviolet-B.

Various natural phenomena can disturb the processes that
continually create and destroy stratospheric ozone and thus increase the level of ultraviolet-B that reaches earth. Large volcanic eruptions, such as those of Mount Saint Helens in Washington State (1980), El Chichon in Mexico (1982), and Mount
Pinatubo in the Philippines (1991), release high into the atmosphere sulfate aerosols and other particulate matter that destroy
ozone.[5] The effects of these and other recent volcanic eruptions
have been transient. Additional natural factors also influence the
levels of ultraviolet-B that reach the earth, such as the well-
known eleven-year solar cycle and El Niño, the quasi-biennial
oscillation in weather patterns.[6]

Man-made chemicals, most notably chlorofluorocarbons
(CFCs), also disturb the ozone layer. Chemists developed CFCs
around 1930 as alternatives to the toxic and flammable refriger-

The ABCs of the Ultraviolet

Photobiologists divide the ultraviolet spectrum into three bands: ultraviolet-C, from 200 to 280 nm; ultraviolet-B, from 280 to 320 nm; and ultraviolet-A, from 320 to 400 nm.° Radiation above 400 nm (violet light) is visible.

Ultraviolet-A was once considered relatively harmless because direct absorption of ultraviolet radiation by DNA was believed to cause skin cancers and DNA absorption declines sharply above 280 nm. (It is more than 10,000-fold lower at 320 nm than at 280 nm).[†] More recent studies, however, indicate that ultraviolet-A can also cause significant health problems.[‡]

One reason that ultraviolet-A is so dangerous is that it penetrates much deeper into the skin than ultraviolet-B.[§] Thus ultraviolet-A can damage cells and connective tissues that are beneath the outermost layer of skin. A consequence of damage to these tissues is the development of tough and leathery skin that eventually becomes wrinkled and aged-looking.

A second reason that ultraviolet-A is dangerous is that it can apparently activate certain endogenous molecules in the skin, called photosensitizers, that can damage DNA, proteins, and other molecules. In fact, many endogenous molecules can potentially serve as photosensitizers, including flavins, pterins, quinones, porphyrins, and steroids. A distinguishing feature of the indirect damage caused by ultraviolet-A is that it seems to require oxygen, suggesting a role for reactive forms of oxygen, such as the toxic singlet oxygen.[#] Indeed, many scientists, believing that reactive forms of oxygen play an important role in causing many cancers and other diseases, strongly recommend such dietary antioxidants as vitamins A, C, and E, to counter such effects.°°

ants, ammonia and sulfur dioxide, that were commonly used at that time.[7] CFCs are inexpensive, easy to produce, nonflammable, nontoxic, noncorrosive, and relatively nonreactive. In addition to serving as coolants in refrigerators and air conditioners, CFCs have also been used as propellants in spray cans, foaming agents, insulation, and cleaning agents for electronic and mechanical equipment.

The stability of CFCs has allowed them to rise up into the stratosphere by convection and turbulent mixing.[8] Once in the stratosphere, CFCs absorb ultraviolet radiation, which causes them to release their chlorine atoms, as demonstrated by F. Sherwood Rowland and Mario Molina in 1974. The atomic chlorine (Cl) and chlorine monoxide (ClO) that result can act as catalysts that destroy ozone.[9] A single chlorine atom destroys about 100,000 molecules of ozone before it precipitates from the stratosphere as HCl.[10] The presence of CFCs in the stratosphere has contributed to a reduction of approximately 5 percent in global ozone levels since 1970.[11] Rowland and Molina, along with Paul Crutzen, were awarded the 1995 Nobel Prize in chemistry for discovering the significance of CFCs in ozone destruction.[12]

In the 1980s atmospheric scientists first recognized a dramatic and unforeseen reduction of stratospheric ozone over Antarctica.[13] This "ozone hole," which occurs in late winter and early spring, is believed to be caused by atmospheric CFCs. Atmospheric scientists have more recently found significant though less severe stratospheric ozone reductions in the Arctic as well.[14]

The discovery of the Antarctic ozone hole led to international protocols in 1987, 1990, and 1992 designed to eliminate production and use of CFCs.[15] Indeed, global use and production of CFCs has declined since the late 1980s. Nevertheless, these compounds have such long lifetimes in the stratosphere that they will remain there for many years, and stratospheric ozone levels will remain below normal for many years. The level of stratospheric ozone was projected to be at a minimum around the year 2000 and is not expected to return to 1970 levels until about the year 2050.[16]

A decline in stratospheric ozone levels, which translates into an increase in terrestrial ultraviolet-B, has been measured at various locations around the world since 1974.[17] Terrestrial measurements of ozone levels can be difficult, since ground-level ultraviolet-B levels depend on various environmental factors including cloud cover, haze levels, and the earth's albedo (surface reflection).[18] In addition, ozone levels change throughout the year. In the Northern Hemisphere, stratospheric ozone levels are maximal from February to May and minimal from September to December; in the Southern Hemisphere, it is just the opposite.[19]

Even though stratospheric ozone levels have declined throughout the world, levels of ultraviolet-B in some urban areas of the United States have actually decreased since 1974.[20] Apparently, pollution from automobiles and industry has increased ozone levels in the troposphere (which lies beneath the stratosphere), and this tropospheric ozone screens out the part of the measured ultraviolet-B that is from scattered radiation.[21] The downside of tropospheric ozone is that direct exposure can be dangerous to plants, animals, and people, in particular by exacerbating such respiratory problems as asthma.[22] In brief, ozone is good in the stratosphere but bad in the troposphere.

It is likely that the increase in ultraviolet-B has had detrimental effects on agricultural yields.[23] The global increase in ultraviolet radiation may also be harmful to many of the earth's natural ecosystems, for if a single species of an ecosystem is disturbed, it may cause reverberations that are felt throughout the ecosystem.[24] Some biologists believe that the increase in global ultraviolet radiation, among other factors, is contributing to the global decline of many amphibian species.[25]

The U.S. Department of Agriculture is responsible for assessing the effects of ultraviolet radiation on agricultural crops and ecosystems.[26] It has established two networks across the country for measuring ultraviolet-B: a research network, which consists of a small number of stations with spectral radiometers for high-resolution measurement of ultraviolet radiation, and a climatological network, which consists of a large number of stations with lower-cost instruments for broadband spectral measurements. These networks provide information via the Internet to the agricultural community and others about the geographical distribution of ultraviolet radiation, the changes in ultraviolet radiation over time, and the climatological and other factors that affect terrestrial levels of ultraviolet radiation.[27]

Ultraviolet Radiation and Human Health

Although the harmful effects of ultraviolet radiation have become common knowledge, absorption of ultraviolet radiation by the skin also has a beneficial effect, because it promotes synthesis of vitamin D, the precursor of a steroid hormone (1,25-dihydroxyvitamin D) necessary for bone growth in children, bone maintenance in adults, and prevention of osteoporosis and fractures in the elderly.[28] Severe vitamin D deficiency can lead to rickets, a disease in which bones and cartilage are incompletely calcified. Cod liver oil, which is rich in vitamin D, was once given to children to prevent rickets.[29] Nowadays, many foods, such as milk and cereals, are fortified with vitamin D, and severe deficiency of this vitamin is rare in developed countries, though many older and institutionalized people still suffer the effects of mild vitamin D deficiency.[30]

Nearly all the other effects of ultraviolet radiation on human health are negative. Chronic exposure to low levels of ultraviolet radiation, for example, can lead to cortical cataracts, a clouding of the eyes' lenses.[31] The vision-blurring effects of cataracts can sometimes be ameliorated by treatment with an ultrasonic probe; in more severe cases, the eyes' lenses are removed and the patient must wear special contact lenses or glasses.[32] It is difficult to predict the effect of ozone depletion on the future incidence of cataracts, because the wavelength sensitivity and dose dependency of cataract formation in humans are not well known.[33]

Skin cancer is the best-known harmful effect of ultraviolet radiation. In fact, exposure to ultraviolet radiation is the most significant cause of skin cancer.[34] About 1.3 million Americans were expected to get some type of skin cancer in the year 2000 (see sidebar, "Epidemiology of Melanoma"). Squamous cell and basal cell carcinomas are the most common forms of skin cancer.[35] Fortunately, these two forms generally are not serious because they do not readily metastasize (spread throughout the body) and are usually detected when they are small and treatable.

Melanoma, skin cancer of the melanocytes (pigment cells), is a more serious form of skin cancer because it often metastasizes.[36] Exposure to ultraviolet radiation early in life, particularly at levels sufficient to cause sunburns, makes people more prone to melanoma than exposure later in life.[37] If estimates prevailed, about 48,000 Americans developed malignant melanoma in the year 2000, and about 7,700 died from it. The increase in sunbathing and the decrease in ozone over the past few decades are responsible for the increasing incidence of skin cancer.[38] It is difficult to forecast the effect of ozone depletion on the future incidence of melanoma, for there is lively debate among scientists about the wavelength sensitivity of melanoma induction in humans.

Ultraviolet radiation causes skin cancer by damaging the DNA of skin cells. It has been estimated that after a sunny day at the beach, a single exposed skin cell suffers 100,000 to 1,000,000 damaged sites in its DNA.[39] If this DNA damage occurs in important growth regulation genes, known as oncogenes or tumor-suppressor genes, then the cells may proliferate uncontrollably into cancer.[40] In fact, damage of the p53 tumor-suppressor gene

Epidemiology of Melanoma

In the United States, melanoma accounts for about 3.5% of all skin cancers (~48,000 cases per year[††]) but about 80% of all skin cancer deaths, or roughly 7,700 people per year.[‡‡] Two other skin cancers, basal cell carcinoma and squamous cell carcinoma, are much more common (~1.3 million cases per year[§§]) but less dangerous because they rarely spread to other parts of the body. From 1950 to 1991 the incidence of melanoma in the United States rose about 3.5-fold, and the mortality rate rose about 1.5-fold.[##] Currently, the number of people who develop melanoma increases by about 4% per year.[***] Most of this increase is attributed to increased exposure to ultraviolet radiation.[†††] The incidence of melanoma among whites is about tenfold higher than among blacks. In men the disease often begins in the trunk, whereas in women it often begins on the lower legs.[‡‡‡] Early detection aids in survival, as more than 90% of those with localized melanoma survive for more than five years, whereas only 15% of those with distant spread of melanoma survive for five years.[§§§]

by ultraviolet radiation is an important initial step in promoting the proliferation of malignant cells in squamous and basal cell carcinomas.[41]

Direct absorption of ultraviolet radiation by DNA is the first step in one major pathway for induction of skin cancer (see figure A3, Appendix). Short-wavelength ultraviolet radiation, the same wavelength region most strongly absorbed by ozone, is most responsible for this effect.[42] Other molecules in the skin, however, absorb significant longer-wavelength ultraviolet radiation, and excitation of these other molecules by light can lead to structural damage of DNA.[43]

In addition, absorption of ultraviolet radiation by certain other molecules in the skin, possibly transurocanic acid, causes a general weakening of the immune system.[44] This effect, termed photo-immunosuppression, weakens the immune system and allows cancer cells to metastasize. Thus patients who take immunosuppressive drugs following an organ transplant are particularly prone to skin cancer because of the additional burden of photo-immunosuppression. Epidemiological studies are needed to investigate the effect of ultraviolet radiation on people who have immune systems weakened by infectious diseases or vaccinations.[45] Such studies are difficult to perform, however, because it is difficult to measure ultraviolet dose rates from natural exposures to sunlight within a population.

DNA Repair

Fortunately, humans and other organisms have special enzymes that can repair damaged DNA. These repair systems are very important in preventing skin cancer.[46] People with *xeroderma pigmentosum*, a rare heritable skin disease characterized by deficient DNA repair systems, are more than one thousand times as susceptible to skin cancers as unaffected people.[47]

One of the most important and interesting DNA repair systems is photoreactivation, in which ultraviolet-A and blue light (360–500 nm) activate photolyase, an enzyme that repairs DNA damage.[48] There are two main types of photolyases: one type repairs the most common form of ultraviolet-induced DNA damage, known as cyclobutane pyrimidine dimers (CPDs), and these enzymes are called CPD photolyases.[49] CPD photolyase has been found in bacteria, single-celled eukaryotes, plants, fungi, fish, and marsupials.[50] *Escherichia coli*, a common bacterium that lives in the human gut, has the enzyme, even though it mostly inhabits an environment totally devoid of light. The second type of photolyase was more recently discovered. It repairs pyrimidine (6-4) pyrimidone sites in damaged DNA, the second–most common form of ultraviolet-induced DNA damage. This enzyme is appropriately called 6-4 photolyase.[51]

Some species, such as *Bacillus subtilis* (a common soil bacterium) and *Schizosaccharomyces pombe* (a type of yeast) appear to lack CPD photolyase.[52] A history of controversy about whether humans and other placental mammals have CPD photolyase is exemplified by two papers published in the *Proceedings of the National Academy of Sciences*.

A group from the University of North Carolina concluded, based on three lines of evidence, that humans lack CPD photolyase.[53] First, ultrasensitive biochemical techniques, which are capable of detecting as few as 10 CPD photolyase molecules per cell, failed to detect any activity in cultured human cells. Second, Southern hybridization tests, which look for similarity in different DNA molecules, showed that human cells do not have DNA sequences that are similar to the DNA sequence of known CPD photolyase genes. Third, northern hybridization tests, which look for the presence of specific RNA molecules, showed that seven human cell types do not express detectable levels of the CPD photolyase gene.[54]

A group at Brookhaven National Laboratory, in contrast, used three different biochemical assays to demonstrate the presence of CPD photolyase in human white blood cells.[55] The levels of photolyase they measured appear to be of physiological significance; in particular, they estimated that sunlight that causes a slight amount of sunburn — a "minimal erythemal dose" — would be expected to repair several hundred CPDs in several minutes. The Brookhaven group gave two reasons why the North Carolina group failed to detect CPD photolyase. First, it used cultured human cells; second, it used cells that were stored at minus 80° C.

(Brookhaven researchers used freshly harvested white blood cells; they also failed to detect photolyase activity when the cells were stored at minus 80° C.)

Why have so many laboratories been unable to detect CPD photolyase in humans? If CPD photolyase is indeed present in human cells, it seems likely that some cells simply do not express this gene. Indeed, expression of this gene can be switched on and off in many different species.[56] Furthermore, the Brookhaven researchers suggest that the failure to detect a human CPD photolyase gene sequence similar to that of other species may be due to the rapid evolutionary change of this gene in placental mammals. This last hypothesis can be tested only by identification of the human CPD photolyase gene, and this has not yet been accomplished.

Sunscreens, Suntans, and Solaria

By now, it is widely known that sunlight can be dangerous and that people who spend time in the sun should use sunscreens. In fact, many sunscreens are effective in protecting against sunburns. The problem is that many sunscreen users believe that they are immune to all the ravages of ultraviolet radiation, so they spend even more time in the sun.

Unfortunately, some sunscreens are not effective in screening out ultraviolet-A radiation. Because ultraviolet-A can cause melanoma and immunosuppression, sunscreen users who spend more time in the sun may be more disease prone.[57] Dermatologists currently recommend using a sunscreen and other forms of protection when you must spend time in the sun, but they advise staying out of direct sunlight as much as possible.[58]

There is little evidence supporting the common belief that getting a tan (which involves synthesis of the tanning pigment, melanin) actually protects the skin by screening out ultraviolet radiation. The major argument in favor of a protective role for a suntan is that racial groups with dark pigmentation have lower rates of skin cancer. This correlation, however, does not necessarily imply that melanin is acting as a sunscreen, and several photobiologists have suggested it has a very different function.[59] Some evidence suggests, in fact, that the DNA damage that occurs following exposure to ultraviolet radiation acts as the molecular signal that triggers melanin synthesis.[60] Although there is controversy about the protective role of melanin and the mechanism for induction of melanin synthesis, dermatologists all agree that there is no way to get a tan without harming the skin.[61]

Solarium operators seem to be unaware of research on the effects of ultraviolet radiation on human health. They claim that tanning in a solarium is safe because the lamps there emit only ultraviolet-A radiation. This claim is doubly false. First, tanning is unsafe whether under natural sunlight, in a solarium, by ultraviolet-B, or by ultraviolet-A.[62] Second, the lights in all solaria emit a small amount of ultraviolet-B. In fact, measurements show that the ultraviolet spectral output of most solaria lights is similar to that of natural sunlight. Ultraviolet-B is one hundred to one thousand times more effective than ultraviolet-A in tanning and skin cancer. Thus the small amount of ultraviolet-B in solaria is disproportionately significant with respect to tanning and skin cancer.[63] Although solaria manufacturers often call their lights ultraviolet-A bulbs, some skin cancer researchers have suggested that calling these ultraviolet-B bulbs would be more accurate.[64]

Many people continue to believe that getting a tan gives them a vibrant and healthy look. But tanning by sunlight or in a solarium is not safe and can cause wrinkling, premature aging of the skin, and melanoma and other forms of skin cancer. To dermatologists and photobiologists, skin that has not been tanned by the sun or a solarium is the most vibrant and healthy looking skin of all.

Suggested Reading

American Academy of Dermatology (www.aad.org).

Cleaver, J. E., and D. L. Mitchell (1997) Ultraviolet radiation carcinogenesis. Pp. 307–18 in *Cancer Medicine*, 4th ed. J. F. Holland et al., eds. New York, Williams and Wilkins.

Madronich, S., et al. (1998) Changes in biologically active radiation reaching the Earth's surface. *Journal of Photochemistry and Photobiology* B 46, 5–19.

Poole, C. M. (1998) *Melanoma: Prevention, Detection, and Treatment.* New Haven, Yale University Press.

USDA UVB Radiation Monitoring Program (uvb.nrel.colostate.edu/).

5 A SAD Tale

A sad tale's best for winter.
— Shakespeare

Physicians have long known that the moods of certain people change with the seasons. More than two thousand years ago, Hippocrates, the father of medicine, reportedly said, "Of constitutions, some are well or ill adapted to summer, others are well or ill adapted to winter." Poets and novelists have also noted these seasonal changes in mood. Emily Dickinson, battered by a moody disposition herself, wrote

> *There's a certain Slant of light,*
> *Winter Afternoons —*
> *That oppresses, like the Heft*
> *Of Cathedral Tunes —*

Victor Hugo, while in exile on the Channel Islands, wrote in *Les Miserables*, "Winter changes into stone the water of heaven and the heart of man."

So Dr. Norman Rosenthal should not have been surprised when he noticed a seasonal pattern in his own mood in the 1970s, after relocating from South Africa to New York City to take a residency in psychiatry.[1] Rosenthal knew that medical professionals before him had noticed seasonal mood changes in certain patients. In fact, seasonal mood changes were widely noted just as psychiatry was evolving into a specialty of medicine in the early 1800s.[2] Rosenthal reports that he remained intrigued by

Corn Flakes and Phototherapy

Although the initial NIH phototherapy studies were groundbreaking, they certainly did not introduce the use of light to treat human disease. At his infamous Battle Creek Sanitarium in Michigan, Dr. John H. Kellogg — brother of Will Keith Kellogg, the creator of Kellogg's Corn Flakes — gave light therapy to between four thousand and five thousand people per year who suffered from diabetes, gangrene, obesity, cirrhosis, and various other ailments.* Like many other of Kellogg's "cures," his light treatments were of dubious value in treating these diseases.

these winter "lows" and summer "highs" as he took a position at the National Institutes of Health in Maryland during the 1980s.[3]

Soon after arriving at the NIH, Rosenthal met Herb Kern, an NIH researcher who became deeply depressed each year with the approach of winter.[4] Kern himself hypothesized that his winter depression was due to a lack of sunlight. Rosenthal, Alfred Lewy, and their colleagues, inspired by Lewy's previous study, which showed that light reduces melatonin levels in humans, hypothesized that abnormally elevated melatonin levels caused Herb Kern's depression. They attempted to alleviate his depression by exposing him to three hours of bright light in the morning and three more hours in the evening over the course of several days (see sidebar, "Corn Flakes and Phototherapy"). Indeed, the NIH researchers noticed a marked improvement in Kern's mood following only three days of light therapy. They published papers describing this interesting phenomenon in 1982 and 1983.[5]

Two years later, Rosenthal and colleagues published the first placebo-controlled crossover study of the therapeutic value of phototherapy.[6] They compared the effect of bright white light of 2500 lux, roughly equivalent to the light level on a cloudy day during late afternoon (see table 1, p. 6), and dim yellow light (500 lux; the placebo) in alleviating the depression of patients who repeatedly suffered from winter depression.[7] Their results showed clearly that bright white light was an effective treatment for winter depression, whereas dim yellow light had no significant effect. In short, phototherapy seemed to act like an antidepressant drug that alters the brain's chemistry. The researchers named the disorder suffered by Kern and similar patients seasonal affective disorder, or SAD.

Psychiatrists now recognize that during the winter, SAD patients experience the onset of a variety of symptoms, including depressed mood, irritability, fatigue, oversleeping, increased appetite, weight gain, and craving for carbohydrates. These symptoms typically fade in the spring, only to return again in the late autumn. Health care providers typically use a self-assessment questionnaire developed by Norman Rosenthal and colleagues for diagnosis of SAD (see table 2). Population surveys with this questionnaire have allowed characterization of the epidemiology of SAD (see sidebar, "Epidemiology of SAD"). One study even reported that a blind patient often developed winter depression and the other hallmark symptoms of SAD.[8] Like other SAD sufferers, this patient responded to bright-light therapy but not dim-light therapy. Apparently, SAD patients with disrupted visual information processing can still respond to phototherapy.

People who think that they suffer from SAD should consult with a physician or mental health professional to confirm the diagnosis. On confirmation a physician will likely recommend the use of a special light box, which consists of a bank of fluorescent tubes that deliver high-intensity white light, for an hour or two each morning. In many cases, phototherapy is a more effective treatment for SAD than psychotherapy or antidepressant drugs. A light box is also much less expensive.

Table 2
Selected questions from the "Seasonal Pattern Assessment Questionnaire," a self-assessment test used for the diagnosis of SAD (N. E. Rosenthal, G. H. Bradt, and T. A. Wehr).

1. To what degree do the following change with the seasons? (Answer as None | Slight | Moderate | Marked | Extremely Marked)
   A. Sleep length
   B. Social activity
   C. Overall mood
   D. Weight
   E. Appetite
   F. Energy level

Where none = 0, slight = 1, moderate = 2, marked = 3, extremely marked = 4. The sum of these six items is the "Global Seasonality Score," which varies from 0 to 24.

2. In the following questions, check for all applicable months. This may be a single month, a cluster of months, or any other grouping. At what time do you . . .
   (Answer as Jan | Feb | Mar | Apr | May | Jun | Jul | Aug | Sep | Oct | Nov | Dec | No particular month)
   A. Feel best
   B. Gain most weight
   C. Socialize most
   D. Sleep least
   E. Eat most
   F. Lose most weight
   G. Socialize least
   H. Feel worst
   I. Eat least
   J. Sleep most

(*Continues*)

Table 2 (*Continued*)

3. Using the scale below, indicate how the following weather
     changes make you feel.

   −3 In very low spirits or markedly slowed down
   −2 Moderately low/slowed down
   −1 Mildly low/slowed down
    0 No effect
   +1 Slightly improves your mood or energy level
   +2 Moderately improves your mood or energy level
   +3 Markedly improves your mood or energy level

   A. Cloudy weather
   B. Hot weather
   C. Humid weather
   D. Sunny days
   E. Dry days
   F. Grey cloudy days
   G. Long days
   H. High pollen count
   I. Foggy smoggy days
   J. Short days

4. By how much does your weight fluctuate during the course
     of the year?

   A. 0–3 lbs
   B. 4–7 lbs
   C. 8–11 lbs
   D. 12–15 lbs
   E. 16–20 lbs
   F. over 20 lbs

5. Approximately how many hours of each 24-hour day do you
     sleep during each season? (including naps)

   Winter (Dec 21–Mar 20)
   Spring (Mar 21–June 20)
   Summer (June 21–Sept 20)
   Fall (Sept 21–Dec 20)

Table 2 (*Continued*)

6. Do you notice a change in food preference during different seasons? If so, please specify.

7. If you experience changes with the seasons, do you feel that these are a problem for you? If yes, is this problem
   Mild | Moderate | Marked | Severe | Disabling

Eye and Brain

How could light possibly affect a person's mood? We still do not have a complete answer, but important clues surely come from looking at the pathway of nerve signals that travel from the retina to the rest of the brain.[9]

Light absorbed by the retina causes an electrochemical signal to travel from the optic nerve to the suprachiasmatic nucleus (SCN) of the hypothalamus. The SCN lies just above the optic chiasm, where certain neural fibers from the left and right eyes cross one another. Nerve fibers spread from the SCN to other regions of the brain, including the pineal gland, a small pinecone-shaped appendage of the brain. The pineal gland of certain fishes, amphibians, and reptiles is itself photosensitive because, like the rhodopsin in our own eyes, it has an opsin that binds to 11-cis retinal (see Chapter 2 and figure A1, Appendix).[10] Just like the rhodopsin in our own eyes, the pineal gland of these distantly related vertebrates is maximally sensitive to blue-green light.[11]

Over evolutionary time, as the skulls of humans and other mammals became thicker, the pineal gland lost its photosensitivity. The mammalian SCN, however, has assumed an important role. Our SCN maintains the body's circadian rhythm, a biological cycle of about twenty-four hours under constant environmental conditions (see Chapter 13). Although the human circadian rhythm is controlled internally by the SCN, it is regulated by

Epidemiology of SAD

Various studies have estimated the prevalence of SAD in the United States at about 2–9 percent.[†] One reason for the different estimates is that different methodologies have been used to define SAD, such as data from a self-assessment questionnaire (see table 2), or self-reported data accompanied by collateral information and/or information from prospective follow-ups.[‡] As might be expected, the prevalence of SAD is lower in southern regions, such as Florida, and higher in northern regions, such as Alaska.[§] SAD is about four times more prevalent in females than males, whereas nonseasonal depression is only about two times more prevalent in females.[#] SAD tends to run in families, and twin studies have shown that there are heritable risk factors.[°°] A significant number of people (~10 percent) suffer from a mild form of SAD, termed "subsyndromal SAD," though its prevalence is difficult to estimate because there is disagreement about exactly how to define this more mild disorder.[††] Phototherapy is also an effective treatment for subsyndromal SAD.[‡‡]

external stimuli, such as light absorbed by the eye.[12] Many researchers have focused on the role of the SCN and circadian rhythm alterations in causing SAD.

At night, electrochemical activity of the SCN causes nerve fibers that are connected to the pineal gland to release norepinepherine (a neurotransmitter), activating receptors on the pineal gland. This increases the activity of N-acetyltransferase, the rate-limiting enzyme for melatonin synthesis, increasing the level of melatonin and decreasing the level of serotonin, a melatonin precursor.[13] During the day, light suppresses the electro-chemical activity of the SCN, so that less norepinepherine is re-leased. This decreases the activity of N-acetyltransferase, lower-ing the level of melatonin and increasing the level of serotonin (see figure A4, Appendix). Many researchers have focused on the role of pineal hormones and neurotransmitters in causing SAD.

Melatonin and SAD

Melatonin was the first hormone to receive attention in SAD re-search. This hormone, which was discovered in 1958, is an in-dole-based compound that is synthesized in the pineal gland from tryptophan, one of the body's ten essential amino acids. During the 1960s and 1970s biologists showed that both diurnal and noc-turnal animals have high melatonin levels at night and low levels during the day. Moreover, biologists showed that melatonin is important in regulating the seasonal changes in body weight, re-production, and physical activity of many animals and the onset of hibernation in hibernating mammals. We now know that many of these effects are mediated by the binding of melatonin to high-affinity melatonin-receptors (termed G-protein coupled recep-tors) that are in the SCN and other regions of the brain and body.[14] Melatonin is also present in diverse species of higher plants and algae, although its physiological role in these organ-isms is still open to speculation.[15]

Early animal studies showed that dim artificial light of ~50 lux, roughly equivalent to the level during early twilight on a clear night, suppresses melatonin levels in many mammals.[16] In addition, early studies with humans demonstrated that melatonin levels are greater at night than day. Until the late 1970s, however, no one had shown a direct effect of light on human melatonin lev-els. At that time, it was believed that melatonin regulation in hu-mans differed from that in other animals.

All this changed in 1980, when Alfred Lewy and colleagues at the National Institutes of Health published a paper showing that bright artificial light suppresses melatonin synthesis in humans.[17] In particular, they demonstrated that a two-hour nighttime exposure to bright artificial light of ~2500 lux strongly reduces human melatonin levels. We now know that the level of light required to suppress melatonin secretion from the pineal gland varies greatly among different animal species. Furthermore, since this early study, more sensitive experimental protocols have shown that light of about 100 lux can cause a statistically significant suppression of human melatonin levels.[18]

An initial hypothesis was that SAD is caused by the abnormal secretion of melatonin or by an abnormal response to melatonin, because melatonin is so important in regulating the rhythms of other animals, and its level is regulated by light.[19] Several lines of evidence indicate that this hypothesis is no longer tenable:

- Light suppresses melatonin levels only when applied early in the morning or late at night, but light is effective against SAD even when applied at midday.
- SAD patients who are treated with light and then given oral melatonin do not experience a complete reversal of symptoms.
- The drug atenolol, which reduces melatonin secretion, is no more effective against SAD than a placebo.[20]

Numerous studies suggest, however, that melatonin has some role in SAD. Experiments with propranolol, a drug related to atenolol, show that it relieves SAD in some patients.[21] Moreover, although oral melatonin does not completely reverse the beneficial effects of light for SAD patients, it does have some effect.[22] There appears to be no simple relationship between melatonin and SAD, but suppression of melatonin by light may be part of the therapeutic mechanism of light. Many psychiatrists recommend that SAD patients avoid taking oral melatonin because it may make them even more lethargic and sleepy, thus exacerbating some of the hallmark symptoms of SAD.

Unfortunately, many unsubstantiated claims have been made about melatonin by the media and those who sell "dietary supplements." Some have claimed that melatonin can not only relieve SAD but also reverse aging, improve your sex life, facilitate weight loss, and cure or prevent cancer, heart disease, Alzheimer's disease, and more. None of these claims is supported by the scientific literature.[23]

Serotonin and SAD

Although there is no simple relationship between melatonin and SAD, it still seems likely that SAD is mediated by an alteration in the levels of certain hormones or neurotransmitters. At present, many researchers are focusing on the serotonin system. Serotonin is an important neurotransmitter that transfers electrical signals between nerve cells in the brain. It is also a metabolic precursor to melatonin, and people tend to have lower levels in the winter than in the summer.[24]

Several lines of evidence suggest an important role for the serotonin system in SAD:

- Many SAD patients are successfully treated with such antidepressant drugs as Prozac, Zoloft, and Paxil (selective serotonin re-uptake inhibitors, or SSRIs), which work by specifically increasing the level of serotonin in nerve synapses of the brain.[25]
- Several studies have shown that SAD patients who are treated with light and then given a diet depleted of tryptophan (a precursor of serotonin) become depressed once again.[26]

- Experiments with meta-chlorophenylpiperazine (m-CPP), a serotonin agonist (a drug that mimics the effects of serotonin), show that it has different effects on SAD patients and normal control patients.[27]

All these studies suggest an important role for the serotonin system in SAD, though the exact nature of its role is still uncertain.

Circadian Rhythms and SAD

Another hypothesis is that SAD is caused by a defect in the body's circadian rhythm, the internal biological clock that is generated endogenously but regulated exogenously (see Chapter 13). In particular, it has been proposed that SAD patients suffer a form of delayed sleep phase syndrome, a sleep disorder in which people have trouble falling asleep at night and trouble waking up in the morning.[28] According to this hypothesis, light exerts its therapeutic effect by advancing the circadian rhythms of SAD patients — that is, by resetting the biological clock to an earlier time.

Several lines of evidence support the circadian rhythm hypothesis of SAD:

- Phototherapy is known to alter human circadian rhythms and is often used to treat jet lag and other sleep disorders.[29]
- Oversleeping is considered one of the hallmark symptoms of SAD.[30]
- Melatonin, the pineal hormone that is affected by light and has been implicated in SAD, is known to affect circadian rhythms.[31]

The circadian rhythm hypothesis of SAD, however, specifically predicts that early-morning phototherapy should be most effective against SAD, that midday light should be ineffective, and that evening phototherapy should be ineffective or even worsen the symptoms. An analysis of studies performed at different centers showed that morning light is more effective, but that midday and evening light were also effective.[32] Some studies, in fact, have shown that the timing of phototherapy is not critical.[33]

In contrast to these phototherapy studies of SAD, other studies have shown that light given in the morning advances the human circadian rhythm, light given at midday has little effect, and light given in the evening delays the rhythm.[34] Moreover, very dim light of ~200 lux, roughly equivalent to the level during sunset on a clear day, can alter the phase of the human circadian rhythm but has no effect on SAD.[35] In fact, dim light of about 200 lux is typically used as a placebo in SAD studies.

Taken in toto, the data indicate that there is no simple relationship between SAD and abnormal circadian rhythm. Many SAD patients appear to have altered circadian rhythms, however, and the antidepressant effect of light may be partly related to phase-advancing of the circadian rhythm.[36]

One reason that scientists have not reached a true consensus on the physiological mechanism of SAD or of phototherapy in treating SAD is that there are no suitable animal models available for studying this disorder. Animal studies have facilitated research of many other diseases, including cardiovascular disease, cancer, and AIDS, but have served as little more than inspirations for SAD researchers.

Another impediment to SAD research is that there is no true placebo for bright-light therapy, so it is difficult to rule out a nonspecific therapeutic effect of light — the placebo effect — in relieving SAD. Most researchers use dim light as the placebo, but patients are certainly aware of the difference between bright light and dim light, so this is far from ideal. Phototherapy, like psychotherapy, sleep deprivation, and electroshock treatment, cannot be concealed from patients, so it is difficult to completely rule out a placebo effect.[37] In fact, one study showed that a special light visor effectively treated SAD, even though it delivered a mere 30 lux of light. The researchers suggest that their light visor functioned as an "elaborate placebo."[38]

Although researchers strive to overcome these and other limitations in their study of SAD, it seems less and less likely that the disparate phenomena associated with SAD and the phototherapeutic treatment of SAD can be explained by a single physiological mechanism. But we do not need to know the mechanism of a cure before knowing that we do indeed have a cure. People who have suffered from seasonal affective disorder in the past need suffer no more because phototherapy really does work. Light really does make right.

Suggested Reading

Oren, D., W. Reich, T. Wehr, and N. Rosenthal (1993) *How to Beat Jet Lag.* New York, Holt.

Rosenthal, N. E. (1993) *Winter Blues. Seasonal Affective Disorder: What It Is and How to Overcome It.* New York, Guilford.

Rosenthal, N. E., and M. C. Blehar (1989) *Seasonal Affective Disorders and Phototherapy.* New York, Guilford.

Shafii, M., and S. L. Shafii (1990) *Biological Rhythms, Mood Disorders, Light Therapy, and the Pineal Gland.* Washington, D.C., American Psychiatric Press.

Society for Light Treatment of Biological Rhythms (www.websciences.org/sltbr/).

6 *The Purple Disease*

Porphyrins red as a summer's rose,
With feathered caps and silken hose,
Pigments of bile, all burnished gold,
Striding their steeds like knights of old.
— *Claude Rimington*

As a resident of upstate New York, where winter days are as short as eight hours and winter snowfall can be up to two hundred inches, I definitely become more upbeat with the arrival of the warm and sunny weather of spring. People with seasonal affective disorder undergo more dramatic changes in mood with seasonal changes in day length. They experience deep depression during the winter and dramatic relief during the spring (see Chapter 5). Of course, nobody — not even people with seasonal affective disorder — should stay out in the sun for prolonged periods, because the sun's ultraviolet rays can cause sunburn and skin cancer (see Chapter 4).

A small number of people have one of the metabolic diseases known as porphyrias, which make them so sensitive to light that even brief exposures to sunlight can be dangerous.[1] The porphyrias (derived from *porphyrus*, the Greek word for purple) are a group of related metabolic diseases in which one of the seven enzymes used to synthesize heme, an important component of hemoglobin, myoglobin, and other proteins, is defective. This condition leads to the abnormal accumulation of porphyrins or porphyrinogens, red or purple colored compounds (see table A1, Appendix).[2] The porphyrins are an important class of biological pigments. Indeed, because chlorophyll is a magnesium-based porphyrin derivative (see Chapter 15), porphyrins are the compounds that make blood red and grass green.

The specific symptoms of the various forms of porphyria differ, but many are accompanied by neuropsychiatric disorders and the excretion of red- or purple-colored porphyrins or porphrinogens in the urine and/or feces. A hallmark symptom of six of the eight forms of porphyria is skin photosensitization, in which areas of the skin exposed to sunlight become inflamed, swollen, and blistered and often exhibit increased hair growth.[3]

Photosensitization

The mechanism of photosensitization was first examined in single-celled organisms around 1900. In one key set of experiments, hematoporphyrin, a mixture of various porphyrin compounds obtained by the chemical degradation of hemoglobin, was added to cultures of *Paramecium,* single-celled, ciliated organisms that have elongated bodies and funnel-shaped grooves that serve as mouths.[4] In darkness, the *Paramecium* cells grew normally, but when exposed to sunlight, they died rapidly.[5] In subsequent experiments, hematoporphyrin was given to mice. The mice survived when kept in darkness, but went into severe shock upon exposure to sunlight.[6] The German scientists who performed these experiments called this response a *Photodynamisagerschein- ung,* a "photodynamic reaction," because it required light, a photosensitizer (in this case, hematoporphyrin), and oxygen.

It only remained to be demonstrated that a similar photodynamic effect would occur in humans. Although the dangers of such experiments were surely evident to early researchers, the German physician F. Meyer-Betz persisted and performed an experiment on himself.[7] On October 14, 1912, he injected himself

Mechanism of Photosensitization

Irradiation of porphyrins (P) with violet light (~400 nm), the region of the spectrum they absorb most strongly, excites them to a higher energy state, called the triplet state (^{3}P).* The excitation energy in triplet state porphyrin is readily transferred to molecular oxygen, which exists as a triplet state molecule ($^3O_2$) in its normally stable ground state.† Transfer of this energy to oxygen causes the formation of singlet oxygen ($^1O_2$), a highly reactive molecule that can chemically alter many important biological molecules, such as amino acids, the building blocks of proteins, and fatty acids, the building blocks of cell membranes. At the physiological level, this leads to inflammation, burning pain, and itching. Overall, the photodynamic reaction of porphyrins can be summarized as:

(1) $P + Light \rightarrow {}^3P$

(2) $^3P + {}^3O_2 \rightarrow P + {}^1O_2$

(3) $^1O_2 \rightarrow \rightarrow \rightarrow$ Cell damage & inflammation‡

with 200 milligrams of hematoporphyrin in solution with dilute sodium hydroxide and physiological saline.[8] During the thirty-minute infusion, he experienced pain in his liver and back and a general feeling of debility. The next day, he exposed himself to sunlight. After about fifteen minutes he developed the classic symptoms of photosensitization on his exposed skin: burning pain, itching, sunburn, and swelling. These symptoms persisted over the following six weeks. The name Meyer-Betz lives on in the scientific literature, but modern scientists surely must wonder whether the physician would have performed his "self-experiment" had he known how much pain he was inviting.

We now know that porphyrins are potent photosensitizers, and we have a basic understanding of the chemical mechanism of the photosensitization reaction (see sidebar, "Mechanism of Photosensitization").[9] Although photodynamic reactions are generally harmful, photodynamic therapy uses photodynamic agents and laser light to selectively kill cancer cells (see sidebar, "Photodynamic Therapy").

Early studies of porphyria were greatly aided by a patient, Mathias Petry.[10] In 1911 physicians noted that Petry suffered from extremely photosensitive skin and that his urine had the color of red wine. They found that Petry's urine was red because it contained copious amounts of porphyrins; similarly, they suspected that accumulation of porphyrins in his skin caused photosensitivity. Because scientists of that era could not easily synthesize or purify porphyrins, Petry's urine provided an abundant source of this compound for research. In fact, the German chemist and physician Hans Fischer reported that he purified 192 milligrams of uroporphyrin esters from 2.4 liters of Petry's urine.[11] Fischer employed Petry as a laboratory aide and used his urine as a source of porphyrins for his groundbreaking studies of porphyrins and tetrapyroles, research that led to his Nobel Prize in 1930. Unfortunately, effective therapy was then unavailable for those suffering from porphyria (see sidebar, "Bloodletting and Other Treatments"), and Mathias Petry died in 1925 at the age of thirty-two.[12] Twenty years later, Hans Fischer committed suicide after Allied forces bombed his research institute in Munich at the close of World War II.[13]

Our understanding of porphyria has increased dramatically since the time of Fischer. By the 1950s the work of David Shemin, Albert Neuberger, Claude Rimington, and others provided a basic understanding of the synthetic pathway of porphyrins.[14] Shemin, like Meyer-Betz, used himself as a guinea pig to study porphyrins; he ingested radioactive chemicals to demon-

Photodynamic Therapy

Photodermatitis occurs in people with any of six forms of porphyria. It may also occur upon contact with certain plants (such as wild parsnip) that contain potent photosensitizing compounds, or upon ingestion of certain photosensitizing drugs.[§] Photodynamic therapy employs the photodynamic reactions of a photosensitizing drug to specifically destroy the cells in cancerous tumors.[#] In photodynamic therapy, a cancer patient is injected with dihematoporphyrin ethers (marketed under the name Photofrin), a mixture of porphyrin-based compounds that is preferentially taken up by tumor cells. When the tumor is irradiated with a laser, the cancer cells are destroyed by the mechanism described in the "Mechanism of Photosensitization" sidebar. A significant side effect of photodynamic therapy is that patients remain very sensitive to light for about six weeks, so they cannot go out into the sunlight following a therapy session. Future research is aimed toward reducing this side effect via photosensitizing drugs that are removed from the body more rapidly and toward developing drugs that have greater affinity for tumor cells. As of the year 2000 photodynamic therapy has been approved in the United States for treatment of obstructing esophageal tumors and certain forms of lung cancer. Clinical trials for treatment of other cancers are under way.[°°]

strate that heme is synthesized by compounds naturally present in the human body. By the mid 1990s, scientists had mapped — identified the chromosomal locations — and cloned all the genes involved in porphyrin synthesis.[15]

Bloodletting and Other Treatments for the Porphyrias

No cures are available for the porphyrias, so physicians can only eliminate precipitating factors and use certain therapies to reduce the severity of acute attacks.

The acute porphyrias. Patients with the acute porphyrias (Doss porphyria, acute intermittent porphyria, coproporphyria, and variegate porphyria) are typically given intravenous heme and glucose and urged to maintain high carbohydrate diets.[††] Heme, which is deficient in patients with these diseases, acts as a negative feedback agent, so its administration inhibits the synthesis and accumulation of porphyrin and porphobilinogen intermediates. Although the acute porphyrias are often accompanied by skin photosensitivity, no treatments have proven effective for this symptom,[‡‡] although canthaxanthin (a beta-carotene analogue) has been used in patients with variegate porphyria.[§§]

Porphyria cutanea tarda. Porphyria cutanea tarda is one of the few medical conditions still treated by bloodletting, or phlebotomy.[##] This treatment reduces iron levels, thus decreasing the synthesis and accumulation of porphyrin and porphobilinogen intermediates.[°°°] Low-dose chloroquine (also used to treat malaria) may also be given.[†††]

Protoporphyria. The acute skin photosensitivity associated with protoporphyria has been successfully treated with high doses of beta-carotene, which apparently works by quenching singlet oxygen (see equation 1 in sidebar, "Mechanism of Photosensitization").[‡‡‡] Beta-carotene protects plants from high levels of sunlight by a similar mechanism (see "Too Much of a Good Thing").

Erythropoietic porphyria. Erythropoietic porphyria is also associated with acute skin photosensitivity. As with protoporphyria, beta-carotene has been used with some success.[§§§] Charcoal treatments have also been used successfully to reduce porphyrin levels.[###]

A Royal Disease?

In 1966 two British psychiatrists, Ida Macalpine and her son, Richard Hunter, suggested that the well-documented bouts of madness suffered by King George III, which eventually led to his confinement at Windsor Castle in the eighteenth century's version of a straitjacket, were caused by acute intermittent porphyria.[16] Because George III ascended the throne of England in 1760 and ruled during the American Revolution, this hypothesis has a bearing on American as well as British history. The King's bouts of madness were previously classified as manic depression (bipolar disorder).[17] To support their diagnosis of porphyria, however, Macalpine and Hunter searched a wide array of historical documents, including forty-seven volumes of manuscripts by Dr. Robert Darling Willis from the British Museum; eight volumes and ten boxes of the Queen's Council Papers from the Lambeth Palace Library; Sir Henry Halford's records of the king's illness from the Royal Archives at Windsor; and the diary of Sir George Baker, president of the Royal College of Physicians and physician to George III. These historical documents showed that the bouts of madness suffered by George III were accompanied by abdominal pain, tachycardia (rapid heart rate), constipation, and — most notably — urine that had the color of red wine.[18]

In 1968, in collaboration with the renowned porphyrin chemist Claude Rimington, Macalpine and Hunter modified their hypothesis by suggesting that the king had variegate porphyria, not acute intermittent porphyria (see table A1 and figure A5, Appendix).[19] One of the main reasons for the change in diagnosis was that they identified descendants of George III who had episodes of skin photosensitivity. They also found evidence of skin photosensitivity in George III. Skin photosensitivity is a characteristic symptom of variegate porphyria but does not occur in acute intermittent porphyria.

The genetics of variegate porphyria (it is inherited as an autosomal dominant — that is, on a chromosome other than a sex chromosome) indicates that this disease should have been present in the descendants of George III. About half of his fifteen children would be expected to have the gene for variegate porphyria. Thus one might think that it would be simple to test the Macalpine and Hunter hypothesis just by looking through historical documents for evidence of porphyria in the king's children. Variegate porphyria, however, has incomplete penetrance, in that only a few of the people who inherit the defective gene actually develop the disease, while many others are latent carriers of a dominant defective gene.[20] A more extensive search through the Royal family was necessary.

Thus Macalpine and colleagues performed an extensive search for the occurrence of porphyria in the royal houses of Stuart, Hanover, and Prussia.[21] They provided evidence for the disease in Mary, Queen of Scots (1542–87), who apparently passed the defective gene to her son, James I (1566–1625). Furthermore, they found evidence suggestive of porphyria in thirteen other family members. Of course, a major obstacle to their research is that the data of several centuries ago are incomplete: the state of medicine was rather primitive, and porphyria was not recognized as a disease until the late nineteenth century.[22]

To further support their hypothesis of variegate porphyria in King George III, Macalpine and colleagues tried to collect data on living members of the royal family. Because family members with variegate porphyria should have urine and feces with high levels of porphyrins, they asked numerous living members of the royal family for urine and stool samples. As might be expected, there was no cooperation. The researchers then asked medical colleagues who treated members of the royal family for the identity of any patients who had symptoms of variegate porphyria. Dr. Alfred Vanotti, a respected authority on porphyria, told of one patient of royal heritage, identified only as "patient A," whom he diagnosed during the 1940s as having porphyria.[23]

More recently, the daughter of patient A also exhibited some symptoms of porphyria. Unfortunately, Dr. Vanotti's medical records are no longer available, and both mother and daughter refused to allow clinical and biochemical tests.[24]

Scientific Controversy

In spite of the enormous volume of historical documents examined by Macalpine and colleagues, some experts disagreed with their hypothesis of porphyria in George III. In particular, Geoffrey Dean, a highly regarded physician and porphyria expert, who had showed previously that the high incidence of variegate porphyria among South African whites is attributable to a seventeenth-century porphyric Dutch settler named Gerrit Jansz, argued that the evidence was insufficient.[25] Dean and another critic argued that most patients with variegate porphyria suffer few symptoms, other than photosensitive skin, unless they take certain drugs that precipitate attacks.[26] They argued that because the precipitating drugs became available only in the twentieth century, the disorder suffered by George III and his relatives could not have been caused by porphyria.

Macalpine countered this argument by noting that a variety of factors other than drugs can precipitate acute attacks in patients with porphyria.[27] Three well-known precipitating factors are ingestion of alcohol, ingestion of small amounts of lead (which may have been present in George III's drinking water), and insufficient consumption of carbohydrates.[28]

Critics have also argued that if Mary, Queen of Scots, had porphyria, then there should be hundreds or thousands of her descendants alive today who would have inherited her defective gene and experienced porphyric attacks after taking modern precipitating drugs, such as barbiturates or sulfanoamides.[29] Macalpine acknowledged that many thousands of people of Royal ancestry may indeed have the disease today, but has not explained why modern medicine has not identified more of them.[30] Some other critics have noted that the urine examinations of the two living descendants were not truly indicative of porphyria and that the clinical evaluations of the two living descendants were incomplete.[31]

Certainly, all parties must agree that diagnosis of porphyria in George III, who died about two hundred years ago, is a formidable task. Of course, if genetic tests were performed on a sample of tissue from George III, or if his living descendants allowed themselves to be thoroughly examined, then an unambiguous diagnosis would be possible.[32]

This scientific controversy, which raged throughout the 1960s, quieted down in recent years, particularly following the deaths of Ida Macalpine in 1974 and Richard Hunter in 1981, the mother-and-son team who originally proposed the Royal porphyria hypothesis. A team of researchers, however, John C. G. Röhl, Martin Warren, and David Hunt, reopened this issue in the 1990s and recently published a book with their findings, *Purple Secret: Genes, "Madness," and the Royal Houses of Europe*.[33] This important book brings the issue of porphyria in George III and the royal family back to life.

The researchers found extensive new corroborative evidence for porphyria in the royal family.[34] This new evidence includes a large collection of private letters written by Princess Charlotte of Prussia (1860–1919), a great-great granddaughter of George III. In these letters to her physician, Charlotte says that she suffered from abdominal pain, skin rashes, and red-brown urine.[35] All of these are symptoms of variegate porphyria. Additional new evidence strongly suggests that Prince William of Gloucester (1941–72), a great-great-great-great grandson of George III, had porphyria. The authors also identify Princess Adelaide of Prussia (1891–1971), a descendant along the Prussian line of George II (father of George III) as the "patient A" referred to by Macalpine and Hunter. With this identification, they were able to confirm variegate porphyria in Princess Adelaide by study of her medical records, which were preserved in the Thuringian State Archives at Meiningen, Germany. Finally, the authors extracted tissue from the corpse of Princess Charlotte and identified a novel mutation in the gene responsible for variegate porphyria (protoporphyrinogen oxidase).[36] This and additional evidence prompted Dr. Geoffrey Dean, one of the most ardent opponents of the royal porphyria hypothesis, to declare that the diagnosis of variegate porphyria in George III may have been correct after all.[37]

George III has long been viewed as the "mad king of England," but he was certainly not the only monarch touched by porphyria or by madness. Indeed, the mental health of monarchs and other leaders has had great influence on world history over the past two thousand years.[38] Perhaps now that the evidence for variegate porphyria in George III is so compelling, we should refer to him not as the "mad king" but as the "photosensitive king."

Suggested Reading

American Porphyria Foundation (www.enterprise.net/apf/).

Dean, G. (1957) Pursuit of a disease. *Scientific American* 196 (3), 133.

Kappas, A. et al. (1995) The porphyrias. Vol. 2, pp. 2103–59 in *The Metabolic and Molecular Bases of Inherited Disease* (C. R. Scriver et al., eds.). New York, McGraw Hill.

Laser Medical Research Foundation (http://209.41.253.5:80/pdt@lmrf).

Macalpine, I., and R. Hunter (1969) Porphyria and King George the Third. *Scientific American* 228 (7), 38–46.

——— (1969) *George III and the Mad Business*. London, Penguin.

Porphyria (www.familyvillage.wisc.edu/lib_porp.htm).

Röhl, J. C. G. et al. (1998) *Purple Secret*. London, Bantam.

Warren, M. J. et al. (1996) The maddening business of King George and porphyria. *Trends in Biochemical Science* 21, 229–34.

7 *A Novel Method of Weed Control*

*I have great faith in a seed. Convince me that
you have a seed there, and I am prepared to
expect wonders.*
— *Henry David Thoreau*

The sun's radiation is the most important energy source for life on earth. Chlorophyll, the best known plant pigment, uses the sun's energy to drive photosynthesis, the conversion of solar energy into biochemical energy. Nearly all other organisms depend upon plants as food, either directly or indirectly, so they also depend upon photosynthesis. Photosynthesis is the best known example of biological energy transduction, the transformation of light energy into chemical energy.

Plants, animals, and bacteria also have photosensory pigments that are important in sensory transduction, not because they harness energy from the sun but because they allow organisms to sense changes in the light environment. This provides the organisms with important information that controls their development, physiology, and behavior. Rhodopsin, the photoreceptive pigment in our eyes (see Chapter 1), is the best known photosensory pigment, but humans are certainly not the only organisms that sense the sun's radiation, and not all organisms are restricted to the wavelengths that humans can sense. Bacteria, plants, and fungi employ a variety of photoreceptive pigments and elaborate mechanisms for sensing and responding to sunlight.

Phytochrome is the most important photosensory pigment

that plants use to regulate their growth and development. Phytochrome plays a role in regulating virtually all stages of plant development, including seed germination, stem growth, flowering, and the onset of senescence. Plant biologists have intensively studied phytochrome and phytochrome-controlled responses since this pigment was discovered in the 1950s.[1]

Recent research on phytochrome-controlled seed germination provides a simple method that farmers could use to reduce weed growth in their fields.[2] Application of this method could greatly reduce our reliance on herbicides and thereby reduce both the cost of farming and the environmental and public health risks associated with herbicide use.

A brief background of light-induced seed germination and phytochrome is needed to fully appreciate the simple beauty of this new method of weed control.

Early Research on Light-Induced Seed Germination

In the 1800s T. W. Engelmann, a brilliant German photobiologist, first determined the effectiveness of different wavelengths of light in photosynthesis, now called an action spectrum.[3] Engelmann was a student of Wilhelm Hofmeister, the German scientist who discovered phototropism in *Phycomyces* (see Chapter 9) while at the University of Heidelberg.[4] In one early set of experiments, Engelmann directed light through a prism, which separated the white light into its component colors, and then directed the different colors of light onto a filamentous green alga that was growing in solution with aerotactic (oxygen-seeking) bacteria. He found that the bacteria accumulated mainly in red and blue light but not in green light, and he correctly concluded that blue and red light were most effective in causing photosynthesis.[5] Engelmann later applied the same basic technique to determine action spectra for a variety of other light-controlled responses in plants and single-celled organisms.[6] His action spectra experiments, although performed more than a century ago, are still featured in modern biology textbooks.

At about the same time, several other German researchers reported that light promotes the germination of many seeds and that some colors of light are more effective than others.[7] None of these seed germination experiments had the wavelength resolution of Engelmann's action spectra and Engelmann never used his prism technique to determine an action spectrum for seed germination. By 1926 Wolfgang Kinzel had identified 930 species of plants with photosensitive seeds.[8]

In 1934 Lewis Flint of the seed testing lab of the U.S. Department of Agriculture in Beltsville, Maryland, set out to determine which wavelengths of light stimulate seed germination.[9] Working with Arlington Fancy lettuce seeds, Flint found that red light promoted germination but that germination was inhibited when blue light was given after red light.[10] Flint soon moved over to the nearby Astrophysical Observatory of the Smithsonian Institution, where he continued his studies of light-induced seed germination with the physicist Edward McAlister.

McAlister constructed a spectrograph, a device for separating radiation into its component wavelengths, which was much larger than the one used by Engelmann.[11] In particular, he directed the light from a streetlamp through a large prism so that the different wavelengths of light were dispersed across the back wall of his laboratory. In a series of studies, Flint and McAlister found that red light (670 nm) was most effective in promoting seed germination, whereas near-infrared light (730 nm) was not at all effective.[12] These studies were confirmed in 1936 by Dieter Meischke of Germany.[13] Meischke, although using glass filters that had much less wavelength resolution than McAlister's spectrograph, found that red light (600–650 nm) was most effective and that far-red light (750 nm) was least effective in promoting seed germination.

These pioneering photobiology studies were not really pursued during the following ten years. Then Sterling Hendricks, a physical chemist at the U.S. Department of Agriculture, set out to determine an action spectrum for the induction of flowering.[14] Hendricks built his own spectrograph in an abandoned wine cellar, using a ten-kilowatt lamp that he scavenged from an old Washington movie theater and other parts that he found in an old Georgetown streetcar. His spectrograph was larger than the one used by McAlister at the Smithsonian Institution, but the total cost was less than fifty dollars.

In the summer of 1945, while most Americans awaited the end of World War II, Hendricks and collaborators published their action spectrum for induction of flowering in Biloxi soybean.[15] Their action spectrum was remarkably similar to the one previously published by Flint and McAlister for the light induction of seed germination. Each had a maximum in the red region (near 670 nm) and showed that light in the far-red region (beyond 720 nm) had no effect.

A few years later, Hendricks and collaborators demonstrated that when lettuce seeds were given a series of brief flashes of red light (670 nm) and far-red light (730 nm), the color of the final light flash determined whether seeds germinated.[16] If it was red, they germinated; if it was far-red, they did not. This remarkable on-off effect was also found to control flowering and, over the following years, a great many other developmental responses in plants (see figure A6, Appendix).[17]

Then Hendricks and his colleagues carried out a series of physiological experiments to investigate the properties of the ubiquitous plant pigment that controlled these responses. Based purely on indirect, physiological experiments carried out over several years, Hendricks and colleagues made several predictions that can be summarized as:

- The reversible response is caused by a single photoreversible pigment.

- The reversibility is caused by photoisomerization, a light-induced structural rearrangement of the pigment.
- The *in vivo* concentration was between 0.1 and 1 micro-Molar.
- The structure of the pigment's chromophore (light-absorbing component) is a linear tetrapyrole, four pyrole groups attached together.
- The pigment is an enzyme.[18]

In 1959 Hendricks and colleagues carried out the first preliminary purification of this ubiquitous plant pigment and named this protein phytochrome, from the Greek meaning "plant pigment."[19] More than forty years after these predictions were made, we know that the first four are correct.[20] The fifth prediction, while likely to be true, has not yet been proven.[21]

Modern Views of Phytochrome

The photoreversible phytochrome responses discovered by Hendricks and colleagues are now classified as low-fluence responses.[22] Exposure to about a tenth of a second of sunlight induces these responses. In the late 1950s a new type of phytochrome reaction was also found to control certain aspects of plant development, such as the synthesis of flavonoids and other plant pigments. These reactions are induced by about ten million times more light than the low-fluence responses, and they are consequently called high-irradiance responses.[23] By the early 1980s Dina Mandoli and Winslow Briggs and other researchers had characterized a third class of phytochrome responses that are induced by about ten-thousand times less light than the low-fluence responses.[24] These are called very low–fluence responses.[25]

Our current understanding of phytochrome is that it allows plants to sense the color, fluence rate, duration, and periodicity of radiation. Thus phytochrome is a photoreceptive pigment that allows plants to sense the light in their environments, albeit in a more rudimentary manner than animals, because plants lack central nervous systems for processing information. Phytochrome research is now one of the most active and competitive areas of research in plant biology. The field is populated by physiologists, biochemists, biophysicists, and — most notably — molecular biologists.

Few modern plant biologists perform research on light-induced seed germination, and such studies are not as fashionable as they were in Hendricks's day. In fact, most plant biologists would probably regard the study of light-induced seed germination as a rather prosaic, if not antediluvian, pursuit. Biologists have known for some time that phytochrome regulates plant development by controlling the differential expression of genes, so the emphasis today is on using modern molecular techniques to dissect the patterns and mechanisms of gene expression.[26]

An important discovery by modern phytochrome researchers is that most plants seem to have a family of different phytochrome genes. In *Arabidopsis thaliana* (mouse-ear cress) a model organism of the mustard family used for studies of plant molecular biology, there are at least five different phytochrome genes, which are designated as *phyA, phyB, phyC, phyD,* and *phyE.*[27] Apparently, some of these phytochrome genes have different functions and sense different aspects of the light environment. Physiological studies with *Arabidopsis thaliana* have shown that PHYA and PHYB, for example, have different roles in controlling seed germination and flowering.[28]

In addition to the many molecular biology studies of phytochrome, which have cost millions of dollars but produced little of practical benefit, plant physiologists and ecologists have documented the light-induced seed germination in thousands of species over the past thirty years. Although light is certainly not the only factor that controls the germination of seeds in nature, ecologists have shown that it is one of the most important factors.[29] Furthermore, ecologists have shown that the seeds of many weed species can survive for many decades buried in the soil.[30] Physiologists have shown that as weed seeds age, while buried in the soil or held at low temperatures in the laboratory, their photosensitivity often increases from the low-fluence response range to the very low–fluence response range. This phenomenon has been well documented for Chinese thornapple *(Datura ferox)*, a noxious weed of South America. The photosensitivity of these seeds increases by about 10,000-fold following burial in a farmer's field for several months.[31]

In direct contrast to weed seeds, the germination of most crop seeds is unaffected by light.[32] Presumably this is because humankind, over the thousands of years of crop plant evolution, has selected seeds that can germinate beneath the soil surface, where they have easy access to water and nutrients yet cannot be eaten by animals. This selection process has inadvertently resulted in crop seeds that are insensitive to light, because the light level beneath the soil surface is insignificant. Another distinction of crop seeds is that they generally cannot survive in the dormant state for a long time, presumably because it is not possible to select for seeds with both high germinability and longevity (see sidebar, "Longevity of Wheat Grains").[33]

Control of Weeds with Herbicides

Like insects, weeds can be noxious pests that drastically reduce farmers' crop yields and profits. Modern farmers have an enormous arsenal of herbicides at their disposal to kill the weeds that grow in their fields. In fact, there has been a dramatic increase in the variety and total amount of chemicals used to control weeds and other pests in the past twenty-five to thirty years. The Weed Science Society of America, a scientific society for weed-control

Longevity of Wheat Grains

Through the ages, many people have come to believe that germinable grains of wheat have been found buried with the mummified pharaohs inside Egyptian pyramids. In fact, this story is nothing but a myth.* It is doubtful that wheat grains could germinate after a century, let alone several millennia. The ancient cereal grains that scientists have found inside the pharaohs' tombs exhibit physiological and morphological degradation and total loss of viability. In particular, barley grains found in the tomb of King Tutankhamen (~1350 B.C.) exhibited extensive carbonization and no viability.†

specialists, publishes a list of about 350 herbicides whose use they approve.[34]

David Pimentel and colleagues have compiled statistics on pesticide use in the United States.[35] Currently the United States uses about one billion pounds of pesticides every year for killing weeds, insects, rodents, and other pests. Herbicides constitute about 60 percent of the pesticides, and most of these are used by farmers. Farmers do not use herbicides on all of their crops. In general, they do not use them on forage crops, such as hay and other livestock foods, but employ them intensively on row crops, such as corn, cotton, and soybeans. In fact, every year farmers use about 250 million pounds of herbicides on their corn fields alone.

Pimentel and colleagues have estimated that the total cost of weed control in the United States is about 4.1 billion dollars per year.[36] Obviously, this represents a significant cost for farm-

ers. In addition to these direct costs, which are rather easy to calculate, many ecologists and public health scientists believe that
herbicide use incurs additional indirect costs, in that the chemicals adversely affect the environment and public health.[37] There
is ongoing controversy about the significance of these indirect
effects, so their associated costs are difficult to calculate.

In the past, most farmers took an approach to weed control
that could be best described as "spray and kill."[38] In other words,
they sprayed on as much herbicide as they could to kill as many
weeds as they could. The more modern approach to weed control
is called "integrated weed management."[39] This approach em-
phasizes managing populations of different species of weeds and
does not seek to eradicate all the weeds, an impossible goal in any
case, according to weed scientists. Integrated weed management
uses knowledge of the entire agricultural system, including the
biology of the weeds, the biology of the crop plants, the biology of
the insects and other animals associated with the weeds and crop
plants, and the ecological interactions of all these species.

One driving force in the development of integrated weed
management has been response to the deleterious effects that
herbicides are believed to have upon the environment and public
health. The National Academy of Sciences has identified ten major pesticides currently used in the United States that may pose a
risk to human health.[40] Alachlor and oxadiazon, two widely used
herbicides, are on this list.[41] According to a 1990 report by the
Office of Technology Assessment, atrazine, the most widely used
herbicide in the United States, has been found in the groundwa-
ter of at least thirteen states, albeit in very low concentrations.[42]
Alachlor is also a common contaminant of ground water.[43]

A second driving force in the development of integrated
weed-management programs has been evolution of herbicide-
resistant weeds (see sidebar, "Resistance, Resurgence, and Replacement"). This is analogous to what occurred in the early
1960s, when many insect species evolved resistance to DDT due

Resistance, Resurgence, and Replacement

Resistance. In *Silent Spring*, Rachel Carson noted that 137 species of insects had evolved resistance to at least one type of pesticide.[‡] She did not mention herbicide resistant weeds, because the first herbicide resistant weed, common groundsel *(Senecio vulgaris)*, was not identified until 1968.[§] With the increasing use of pesticides over time, there are now at least 273 species of weeds, 504 species of insects and mites, and 150 plant pathogens with pesticide resistance.[#]

Resurgence. In addition to resistance, an evolutionary phenomenon, pesticide use may also lead to resurgence, a rapid population increase, of the remaining pests. This may occur because the few pests that remain following pesticide treatment experience less intraspecific competition for food and other resources and therefore reproduce in much greater numbers.[°°]

Replacement. Use of pesticides may also cause a secondary pest species to replace the primary pest species. This may occur when the pesticide kills off one species (the primary pest) that normally preys upon another species (the secondary pest). Because the predator of the secondary pest has been eliminated, the population of the secondary pest may dramatically increase.[††]

to overuse of this insecticide. In brief, herbicide resistance evolves when certain rare individuals in a population that are resistant to an herbicide preferentially multiply as the herbicide is used repeatedly. There are now at least 273 weed species that exhibit some form of herbicide resistance.[44]

The evolutionary selection for herbicide resistance has

been well documented in many species, particularly in certain populations of a noxious weed called annual ryegrass or rigid ryegrass *(Lolium rigidum)*. Certain populations of this weed have evolved resistance to more than twenty herbicides.[45] Annual ryegrass and many other weeds have also evolved cross-resistance to herbicides, wherein exposure to one herbicide has led to resistance to others.[46] In this case, simply switching to a new herbicide will not work. It seems likely that the relatively common use of combinations of herbicides has led to the evolution of herbicide cross-resistance.

Because weeds can evolve resistance and cross-resistance to herbicides, any decision that a farmer makes about the use of a herbicide affects more than just a single crop from a single season. The choice to use a single herbicide, several herbicides, or no herbicides at all is a decision that a farmer must live with for many growing seasons. It may be possible to determine an allowable weed population size, where the value of the crop gained by eliminating weeds equals the cost of weed control. The consequences of herbicide use, however, are often so complex and so long-lasting that such calculations often have little value for more than one or two seasons. This is one reason why the science of weed control is such a challenging discipline.

Control of Weeds by Plowing at Night

Many millennia before the invention of herbicides, farmers simply plowed their fields to control weeds. Even today, plowing can constitute a valuable part of an integrated weed-management program. Although plowing kills standing weeds, farmers have long known that it often leads to the emergence of new weed seedlings in a few weeks.

Ecologists have shown that a farmer's field can have 50,000 or more weed seeds per square meter buried beneath the soil surface.[47] Plant physiologists have shown that seeds buried more than about one centimeter below the soil surface do not receive enough light to germinate.[48] Do the blades of a plow, which can reach more than a foot beneath the soil surface, bring some of these buried seeds to the surface where their germination is induced by exposure to sunlight?

Two ecologists, Jonathan Sauer and Gwendolyn Struik, began to study this question in the 1960s.[49] In a relatively simple experiment, they went to ten different habitats in Wisconsin during the night and collected pairs of soil samples. They stirred up the soil in one sample of each pair in the light and stirred up the other sample of each pair in the dark. Then they exposed all ten pairs to natural sunlight in a greenhouse. For nine of the ten pairs of soil samples, weed growth was greater in the samples stirred up in light. They concluded that soil disturbance gives weed seeds a "light break," and this stimulates their germination.[50]

More recently, Karl Hartmann of Erlangen University in Germany reasoned that when farmers plowed their fields during the day, the buried weed seeds are briefly exposed to sunlight as the soil is turned over, and that this stimulates their germination.[51] Although the light exposures from plowing may be less than one millisecond, that can be enough to induce seed germination via the very low–fluence response (see table 1, p. 6). Thus the germination of weed seeds would be minimized if farmers simply plowed their fields during the night, when the photon fluence rate is below 10^{15} photons per square meter per second. Although even under these conditions hundreds of millions of photons strike each square millimeter of ground each second, this illumination is below the threshold needed to stimulate the germination of most seeds.

Hartmann says that he was very skeptical when he first came up with this idea because he assumed that such a simple method of weed control as plowing at nighttime must be ineffective or it would have been discovered long ago.[52] But the subsequent experiments, first presented at a 1989 scientific meeting in

Freiburg, Germany, clearly demonstrated that the method can be effective.[53]

Hartmann tested his idea by plowing two agricultural strips near Altershausen, Germany. The farmer Karl Seydel cultivated one strip, repeated threefold, at around midday and the other strip at night. No crops were planted in these pilot experiments, to avoid possible competition with the emerging weeds. The results were dramatic.[54] More than 80 percent of the surface of the field plowed in daylight was covered by weeds, whereas only about 2 percent of the field plowed at night was covered by weeds.

This method of weed control is currently being used by several farmers in Germany. Because many of the same weed species that invade farmers' fields in Germany also invade fields elsewhere in the world, this method should be successful elsewhere. In fact, recent studies at universities in Nebraska, Oregon, Minnesota, Denmark, Sweden, and Argentina support this idea.[55]

This new method of weed control may play a valuable part of integrated weed control programs of the future. Like all weed control methods, however, it has several limitations:

- The number of buried seeds in most farmers' fields is so great and weed seeds remain viable for such a long time (many decades) that many viable weed seeds will remain in the soil, because nighttime plowing does not kill unexposed seeds. A partial solution to this problem would be for the farmer to initially plow his field during the daytime, to promote the germination of weed seeds and deplete the underground seed bank. Following this initial daytime plowing, the farmer could use nighttime plowing to reduce weed growth.
- There is a technical limitation of nighttime plowing that farmers must overcome. Any light seen by a hu-

man being will also be detected by a seed. Thus the farmer's tractor must be equipped with special infrared headlights, and the farmer himself must wear infrared-sensitive glasses, of the type used by soldiers during nighttime military maneuvers.

- Another limitation is that exclusive reliance upon nighttime plowing for weed control can lead to the dominance of weeds whose seeds germinate in darkness. This is analogous to what happens when herbicides are used to control weeds. Individuals in the population that are resistant to the chemical preferentially reproduce and eventually dominate. Hartmann suggests that several different methods of weed control be alternated, so that selection for light-independent germination does not occur.[56]

Obviously, nighttime plowing is no panacea — it will not instantly eliminate the need for all herbicides. But it may reduce our dependence on herbicides and may play a valuable part in integrated weed-management programs of the future. Exactly how nighttime plowing should be incorporated within an integrated weed management program remains to be determined by future research.

Why did it take so long to discover this simple and effective method of weed control? Perhaps the hundreds of scientists involved in phytochrome research do not have a mind-set that prepares them for a simple, low-tech solution. On the other hand, perhaps this novel method of weed control is not so new after all.

Eric Sloan, a well-known chronicler of early Americana, writes in his book *Diary of an Early American Boy,* "It is interesting to note here how many farmers used to work in what we now call darkness. . . . Farmers frequently did their haying at night, using the moon or the stars for illumination, and taking advantage of the coolness of summer night."[57] Modern farmers may be able to use a few simple modifications of the practices of their forefathers as part of a modern approach to weed control.

Suggested Reading

Hartmann, K. M., and W. Nezadal (1990) Photocontrol of weeds without herbicides. *Naturwissenschaften 77*: 158–63.

Sage, L. C. (1992) *Pigment of the Imagination: A History of Phytochrome Research.* San Diego, Academic Press.

Steingraber, S. (1997) *Living Downstream.* New York, Addison-Wesley.

Weed Science Society of America (ext.agn.uiuc.edu/wssa/).

8 Light and Beer

John Barleycorn was a hero bold,

Of noble enterprise,

For if you do but taste his blood,

'Twill make your courage rise.

— *Robert Burns*

Have you ever cleaned and dried your favorite beer glass, chilled your favorite brew, popped the cap and filled your glass, only to be put off by a "skunky" odor?

If so, your beer was probably destroyed by light in the well-known sunstruck reaction (see figure A7, Appendix, and sidebar, "A Windowsill Experiment"). In this reaction, ultraviolet or blue light chemically transforms certain hop compounds in beer and gives the beer an unpleasant odor.[1] This is why brewers must always keep their beer away from light. Some commercial brewers use an alternative approach to protect their beer. They brew with a mixture of chemically modified hop compounds that are stable in the light so that they can sell beer packaged in clear bottles.

The sunstruck reaction is certainly the most familiar light-promoted chemical reaction in beer. But more than four hundred chemicals have been identified in beer, in addition to water and ethanol, so it is likely that light promotes other chemical reactions that simply have not yet been identified.[2] Moreover, as we shall see, brewers can use ultraviolet radiation to improve the utilization of hop bitter acids and to synthesize novel bittering compounds from hops.

A Windowsill Experiment

What does sunstruck beer taste like? A simple way to find out is to do an experiment on your own windowsill at home. Set two identical bottles of beer on a sunny windowsill. One bottle should be fully exposed to sunlight; the other bottle should be completely shielded from sunlight but exposed to the same vicissitudes of temperature. It is best to choose light-colored beers that are highly hopped and come in green or clear bottles. Let the two beers sit for several hours on your sunny windowsill and then do a side-by-side taste test (or smell test). The beer exposed to sunlight may have a skunky odor due to the presence of prenylmercaptan. You may not be able to tell the difference if the beers were already sunstruck or if the brewery used photostable hop compounds.

Hop Alpha Acids and Beta Acids

It is necessary to have some understanding of the alpha acids and beta acids of hops to understand the sunstruck effect. These are two classes of related resins that occur naturally in hop flowers, the part of the plant used for making beer.[3] Their chemical structures are similar, but alpha acids are generally more abundant and generally contribute more to beer flavor.[4] The amounts of alpha and beta acids vary according to the hop variety (Cascade, Saaz, Golding, Northern Brewer, and so on) and the environment in which the hop plant is grown.[5] This is why brewers must rely upon a chemical analysis of the particular harvest of hops that they are using in their beer.

The three main types of alpha acids in hops (humulone, typically 35–70 percent; cohumulone, typically 20–65 percent;

adhumulone, typically 10–15 percent) have similar chemical structures and chemical reactivities. Likewise, the three main types of beta acids (lupulone, typically 30–55 percent; colupulone, typically 20–55 percent; adlupulone, typically 10–15 percent) have similar structures and chemical reactivities.[6]

Plant biologists refer to alpha acids and beta acids as phytochemicals or "secondary compounds," because they are not absolutely necessary for survival of the hop plant (unlike DNA, RNA, proteins, and lipids).[7] Many secondary compounds appear to ward off attack by insects and other predators, and plant biologists believe that natural selection has favored plants that make secondary compounds that increase their chance of survival and reproduction.[8] Another possible reason for the evolution of alpha and beta acids may be that these compounds absorb ultraviolet radiation and could act as natural sunscreens for the plant. Because ultraviolet radiation can damage DNA (see Chapter 4), there are strong selection pressures for plants to reduce such damage. In fact, other plant pigments, such as flavonoids, have been shown to act as natural sunscreens that prevent DNA damage.[9]

The ultimate reason why hop plants evolved the biochemical pathways for production of alpha and beta acids may never be known. Over the past several hundred years, hop growers have artificially selected many varieties of hops that make alpha acids, beta acids, and other compounds that they find desirable for making beer, not for survival of the plant in nature.

Brewing and Transforming Alpha Acids

Brewing consists of three stages: the mash, the wort boil, and fermentation.[10] During the mash, the brewer soaks malted (partially germinated) grains in warm water to enzymatically transform the starch into sweet sugars that can be digested by brewing yeast. Next, the brewer removes the wort (sweet liquid) from the spent grains and boils it for an hour or more, with specially timed additions of hops. Following the boil, the brewer cools the wort and adds yeast for fermentation of the sugars into ethanol.

During the wort boil, the alpha acids of hops are isomerized (chemically changed) into iso-alpha acids, bitter-tasting compounds that balance the residual sweetness in the fermented beer.[11] Unisomerized alpha acids have little or no taste.[12] In addition to providing bitterness, iso-alpha acids stabilize beer foam and inhibit bacterial growth.[13] Isomerization occurs optimally in mildly alkaline solutions (pH about 10–11).[14] Because beer wort is mildly acidic (pH about 5–6), and alpha acids are not very soluble in beer wort, conversion into iso-alpha acids during the wort boil is incomplete, about 45 percent following a typical one-hour wort boil.[15] The concentration of iso-alpha acids in beer varies according to beer style but generally lies between about ten parts per million (or 10 IBUs, international bittering units) for an American light lager, such as Budweiser, and about sixty parts per million for an India pale ale, an especially hoppy style originally brewed in England for export to India.[16] Many styles of beer have other substances that also contribute to bitterness, such as various compounds in the dark specialty malts used in stouts and porters.[17]

Although alpha acids are generally more abundant and more important in contributing to beer bitterness, beta acids also contribute. In particular, as hops age, alpha acids and beta acids become oxidized (chemically combined with oxygen). Oxidized alpha acids cannot be isomerized into iso-alpha acids during the wort boil, so the hops have a reduced bittering potential.[18] Oxidized beta acids, however, are themselves bitter-tasting compounds that are soluble in beer, so their presence may compensate for the reduced formation of iso-alpha acids.[19] Unfortunately, oxidized beta acids have a flavor profile different from iso-alpha acids, so most brewers consider their presence undesirable.[20]

This is one reason why brewers always try to use fresh hops and store their hops in oxygen-free containers.

The Sunstruck Reaction

The sunstruck reaction is a photochemical (light-promoted) reaction of iso-alpha acids that can occur whenever beer is exposed to light. It can take place at any stage after the formation of iso-alpha acids during the wort boil, so brewers must keep their boiled wort, fermenters, and bottles away from light. The key reaction is the photolysis (splitting apart by light) of an iso-alpha acid and combination of one fragment with sulfur to form a foul-smelling compound called prenyl-mercaptan (see sidebar, "Chemistry of the Sunstruck Reaction"). Other foul-smelling chemicals (methanethiol, methional) may also be formed in beer following exposure to light, but their contribution to beer skunkiness apparently is relatively small. The sunstruck reaction does not occur when copper ions are added to beer.[21] Although the mechanism of this prophylactic effect is unclear, it suggests that using a copper brewing vessel may protect beer from the sunstruck reaction.

Mercaptans are notorious because in addition to causing beer skunkiness, they contribute to the natural odors of skunks, garlic, rotting chicken, bad breath, and many other natural malodors.[22] Similar sunstruck phenomena occur in beverages that do not have any hop compounds, such as white wine, champagne, milk, and rice vinegar, although the specific skunky odor of sunstruck beer is unique.[23]

Unfortunately, the presence of a minuscule amount of prenyl-mercaptan can be disastrous for a bottle of beer. While iso-alpha acids occur at about ten to sixty parts per million in beer, many people can taste prenyl-mercaptan at a level approaching one part per billion in beer.[24] That's something like being able to taste one drop in about 13,000 gallons of beer.[25]

Chemistry of the Sunstruck Reaction

In the sunstruck reaction, ultraviolet radiation splits off the five-carbon chain attached to carbon-4 of the five-carbon iso-humulone ring, a Norrish type I reaction (see figure A7, Appendix).[*] This five-carbon chain (3-methyl-2-butenyl radical) combines with a thiol radical (-SH) to form the very foul-smelling prenyl-mercaptan (3-methyl-2-butene-1-thiol). The thiol radical presumably comes from a free cysteine (a sulfur-containing amino acid) or from a cysteine attached to a polypeptide or protein. Several additional products are also formed in this photoreaction (dehydrohumulinic acid and cis-iso-humulone). Riboflavin (vitamin B-2) has been strongly implicated in the visible light induction of the sunstruck effect, and it has been proposed that flavins, which strongly absorb blue light, pass their excitation energy to iso-alpha acids and promote their photolysis.[†] The excited state of riboflavin (triplet energy, 21.5 kcal mol^{-1}), however, is significantly lower in energy than the excited state of iso-alpha acids (triplet energy, 72.5 kcal mol^{-1}), making this reaction energetically impossible.[‡] Clearly, the mechanism of visible light promotion of the sunstruck reaction is not completely understood.

Iso-alpha acids principally absorb radiation in the ultraviolet region of the spectrum, which is invisible to humans.[26] But blue light also promotes formation of prenyl-mercaptan in beer.[27] Apparently, blue light–absorbing molecules in beer promote the photolysis of iso-alpha acids into prenyl-mercaptan. The concentration of riboflavin (vitamin B-2) in beer correlates with the promotion of prenyl-mercaptan formation by blue light.[28] Because flavins absorb blue light and occur in beer at a level of several hundred parts per billion, it is likely that they promote the sunstruck reaction in visible light.[29] Thus reducing the flavin content of beer should also reduce the sunstruck reaction. Unfortunately, there is no simple method that brewers can use to remove the small amount of flavins present in their beer.

Green and brown bottles screen out ultraviolet radiation and prevent iso-alpha acids from absorbing ultraviolet radiation. Brown bottles are better than green bottles in preventing the sunstruck reaction, because blue light easily penetrates through green glass but not brown. Some initial data on beer exposed to sunlight indicates that prenyl-mercaptan formation is about twice as high in clear bottles as in green bottles, and about six times as high in green bottles as in brown bottles.[30]

The sunstruck reaction is a greater problem for light-colored beers than dark-colored beers.[31] This is apparently because light penetration through a dark beer is very poor, and little of the light can be absorbed by flavins or iso-alpha acids. Moreover, any skunkiness present in a dark beer, such as a porter or a stout, may be masked by the stronger flavors of those beers.

A related issue is the storage of beer. It is usually no problem for a beer drinker to store beer in a cool dark basement or refrigerator, and indeed these are ideal places. But what about a beer vendor who wants to display his beer? Incandescent lights, which have low output in the ultraviolet and blue regions of the spectrum, are suitable, but they are often more expensive to run than fluorescent lights.[32] Thus many vendors display their beer under "warm white" fluorescent lamps, which have low output in the blue and ultraviolet regions. "Daylight" and "cool white" fluorescent lamps have high output of blue and ultraviolet radiation, the same spectral regions that promote the sunstruck reaction.[33]

All commercial brewers are aware of the sunstruck reaction, but many continue to use green or clear glass bottles to satisfy consumer preferences. Indeed, a number of commercially available beers are sunstruck. Many large breweries, however, use chemically modified iso-alpha acids that are stable in the light (see figure A8, Appendix). Kalsec, a company based in Kalamazoo, Michigan, prepares many different types of photostable hop compounds that are not degraded by light.[34] It sells these products, the compositions of which are custom-designed and proprietary, to six of the seven largest breweries in the world and to numerous microbreweries. Some beer aficionados do not like these modified hop compounds, claiming that they have bitterness profiles that differ from natural iso-alpha acids.

Ultraviolet radiation can also be used to convert alpha acids into iso-alpha acids. Irradiation of humulone under the proper conditions, for example, gives pure trans-iso-humulone, a bitter-tasting iso-alpha acid.[35] Hop chemists often use this photochemical reaction to prepare trans-iso-humulone in the laboratory because it is a relatively simple technique for making pure samples. Trans-iso-humulone is often used as a chemical standard for bitterness in taste tests.[36]

The yield from the photochemical transformation of humulone into trans-iso-humulone is about 90 percent, much higher than the yield of iso-alpha acids during the wort boil.[37] Breweries wanting to improve the efficiency of hop utilization may be able to scale up this photochemical reaction by incorporating a "hop photoreactor," a device that promotes alpha-acid isomerization, within their brewing operation.[38]

Photochemical Conversion of Beta Acids to Alpha Acids

Ultraviolet radiation can also convert beta acids into alpha acids.[39] When the beta-acid lupulone, for example, is given ultraviolet radiation under the proper conditions, it is converted to 4-deoxy-humulone, in a reaction chemists call a Norrish type II elimination. The 4-deoxy-humulone can then be easily oxidized to give humulone, an alpha acid.

Because hop beta acids can be oxidized into compounds that impart a bad taste to beer, breweries could conceivably use a scaled-up version of this photochemical reaction to improve the shelf life of their hops. Furthermore, because alpha acids are precursors of the iso-alpha acids, a scaled-up application of this photochemical reaction might allow more efficient extraction of bitter acids from hops. The main limitation preventing application of this method is that the yield from this photochemical reaction is very low.[40]

Considering the great panoply of chemicals and chemical reactions that occur in beer, it may seem surprising to some people that beer tastes as good as it does. But when considering the flavors of different beers, one must account for the physiology of taste as well as the chemistry of beer, for many of the chemical compounds in beer are below the taste threshold. Prenylmercaptan, the product of the sunstruck reaction, is the rare compound that we can taste at a concentration of about one part per billion.

Suggested Reading

De Keukeleire, D. (1991) Photochemistry of Beer. *The Spectrum* 4 (2): 1, 3–7. Center for Photochemical Sciences, Bowling Green State University.

Jackson, M. (1993) *Beer Companion.* Philadelphia, Running Press.

Peacock, V. (1997) Fundamentals of Hop Chemistry. *Malting and Brewing Association of America Technical Quarterly* 34, 4–8.

Templar, J., et al. (1995) Formation, measurement, and significance of lightstruck flavor in beer: A review. *Brewers Digest,* May, pp. 18–26.

Verzele, M., and D. De Keukeleire (1991) *Chemistry and Analysis of Hop and Beer Bitter Acids.* New York, Elsevier.

Nishek, D. (ed.) (1997) *Zymurgy: The Classic Guide to Hops.* Special Issue. 20 (4).

9 Phycomyces, *the Fungus That Sees*

The pipette is my clarinet.

—*Max Delbrück*

Most of us would consider vision the most important of our five senses, for it provides us with so much important information about the world. In fact, our eyes seem remarkably well adapted, in an evolutionary sense, to the light environment of the earth. The sunlight that reaches the earth is mostly between 400 nm (violet) and 700 nm (deep red) and our visual system is maximally sensitive to radiation in this part of the spectrum.[1] Our eyes are so sensitive to light that a single rhodopsin-containing rod cell of the retina can respond to the absorption of a single photon, and we can sense a flash of light when only about six different nearby rod cells absorb photons (see Chapter 1).

Another remarkable feature of the human visual system is that it adapts, in a physiological sense, to an enormous range of ambient light levels. The level of light on the earth varies by about seven or eight orders of magnitude over the course of a day (see table 1, p. 6). Physiological adaptation processes allow the human visual system to adjust its sensitivity so that it can respond to light levels that vary by about 10^{12}-fold, over a range that covers that found on the earth's surface.[2] Feedback pathways that connect the different steps of the visual response are responsible for physiological adaptation to ambient light over such an enormous range.[3] Indeed, feedback pathways regulate adaptation in the sensory systems of all our five senses, as well as in all the various sensory systems of plants and microorganisms.[4]

Although physiological adaptation has been a central theme of research in vision science, it has been neglected in research on the photosensory systems of plants and fungi.[5] Studies of physiological adaptation in the photosensory systems of plants and fungi may reveal important general principles of the feedback pathways that regulate all sensory systems, just as studies of the genetics of yeast, fruit flies, and bacteria have uncovered general principles of genetics that apply to all organisms, including humans.

One fascinating organism in which photosensory adaptation has been intensively studied is *Phycomyces blakesleeanus,* a tiny, Zygomycete fungus whose spore-bearing sporangiophore (stalk) bends toward the light (see figures 5 and 6).[6] *Phycomyces* is closely related to *Pilobolus,* known as the "hat thrower" fungus, which forcibly shoots its spores toward the light. Like many other fungi, *Phycomyces* grows on animal droppings in the wild.[7] Despite this unseemly habit, it has a photosensory system that controls growth and phototropism with ranges of photosensitivity and adaptation comparable to the visual system of humans. In fact, after spending several years of my life studying *Phycomyces* in the laboratory, I saw many similarities between the way it senses light and the way we humans sense light with our eyes — hence the title of this essay. Of course, this primitive fungus cannot really see, for true vision requires a central nervous system.

Perhaps because *Phycomyces* is so small, very few of the many biologists who have studied it have ever seen it in nature. In fact, when I first started working with *Phycomyces* many years ago, I had never seen it in the wild. Thus one Friday evening in May of 1988, I swallowed several million of its microscopic spores. Early morning on the next two days, I rode my bicycle to a nearby state park to deposit my droppings. Soon after, I found hundreds of the little spore-bearing sporangiophores of *Phycomyces* popping up at my special spot. Perhaps future *Phycomyces* researchers of upstate New York now have a better chance of seeing their subject in its natural habitat.

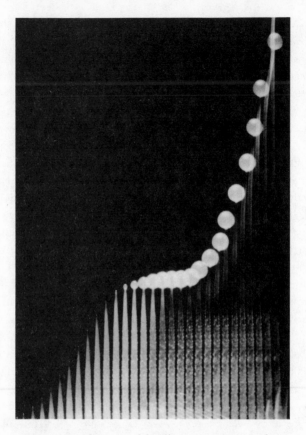

Figure 5. Development of the *Phycomyces* sporangiophore. Photos of the development of a single sporangiophore were taken at one-hour intervals. At left, the sporangiophore is in stage I and has not yet formed a sporangium. At right, the sporangiophore is in stage IVb and has a fully formed sporangium. Most biologists use sporangiophores in stage IVb for experiments. Photo courtesy of David S. Dennison, Dartmouth College, Hanover, N.H.

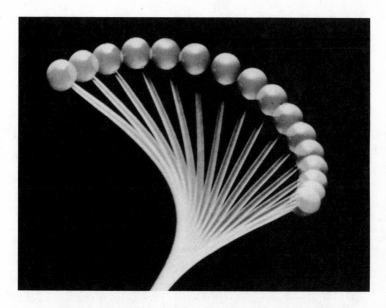

Figure 6. Phototropism in the *Phycomyces* sporangiophore.
A sporangiophore was illuminated with blue light from the side, and
photos were taken at three-minute intervals. The bending rate is about
three degrees per minute. Photo courtesy of David S. Dennison,
Dartmouth College, Hanover, N.H.

In the laboratory, biologists typically grow *Phycomyces* on a
mixture of agar and mashed potatoes.[8] In nature or in the laboratory, the germinating spore of this little fungus produces a slender reproductive organ called the sporangiophore. The sporangiophore grows to several inches in length, although it is only as
thick as a human hair. It is capped by a spherical sporangium that
contains about 100,000 spores, although it is only as large as the
head of a pin. When sown under proper conditions, each spore
gives rise to its own sporangiophore capped by its own sporangium with another 100,000 spores.[9]

The *Phycomyces* sporangiophore alters its rate and direction of growth in response to light, gravity, wind, odors, and the presence of nearby objects.[10] The responses to light have been studied most intensively. The two principal responses are phototropism, bending in response to a directional light source, and the light-growth response, a transient increase in growth rate following an increase in the level of ambient light. Because *Phycomyces* is virtually never seen in nature, we do not know the significance of these and its other sensory responses for survival of individuals in nature.

Early Phycomyces *Research*

The response of the sporangiophore to light was first noted by Wilhelm Hofmeister, a nineteenth century German professor whose scientific genius has been very much overlooked.[11] Hofmeister's observation, recorded in his 1867 treatise *Die Lehre von der Pflanzenzelle*, predates the phototropism experiments performed by Charles Darwin with his son Francis. The Darwins' research on phototropism, however, published in 1880 in *The Power of Movement in Plants*, was much more influential.[12] In fact, because of this book, Charles Darwin would be well known to modern biologists even if he had not written a word on evolution. The Darwins correctly postulated that an asymmetric light field leads to the asymmetric redistribution of a chemical "influence," subsequently identified as indole acetic acid, and this causes phototropism. Although the Darwins made no mention of *Phycomyces* in their book, their experiments certainly influenced all future *Phycomyces* researchers.

Following Darwin's book, *Phycomyces* research was performed by Johann Bruder in Germany, Jean Massart in Belgium, and Edward Castle in the United States.[13] I have always been personally intrigued by Castle's *Phycomyces* experiments, which reveal a truly creative intellect at work. His publications

Edward Castle

In a 1940 paper Edward Castle described a special apparatus (based on a rotating cylinder of film, with *Phycomyces* in the middle) that he built at Harvard University to measure sporangiophore growth rate at 0.2-second intervals, a resolution greater than any of the modern instruments used by *Phycomyces* labs in America and Germany.° Castle reasoned that he should examine sporangiophore growth in a constant environment before he could truly understand sporangiophore responses to light and other stimuli. He showed that the growth rate of the sporangiophore fluctuates in a constant environment and then proposed a model to account for this ever-changing physiological state of the sporangiophore. Perhaps as testament to Castle's insight, the general features of his model still seem plausible.

from fifty or more years ago provide detailed descriptions of the laboratory equipment he built to study the effect of light on sporangiophore growth and the mathematical models he developed to explain his results. One of his papers so fascinated me that it stimulated some of my own *Phycomyces* studies (see sidebar, "Edward Castle").[14]

In collaboration with Ed Lipson and Michael Vinson of the physics department at Syracuse University, I reinvestigated Castle's finding that sporangiophore growth fluctuates in a constant environment.[15] Like Castle, we found that the growth rate fluctuates dramatically when the sporangiophore is held in constant darkness or constant illumination. Furthermore, we found that these growth fluctuations have what physicists call a "deterministic nonlinearity."[16] This deterministic nonlinearity may allow the sporangiophore to better adjust its sensitivity to sensory input so that it can better respond to small changes in the level of light or other sensory stimuli by altering its growth rate.

As an extension of this research, we also studied the growth fluctuations of various behavioral mutants of *Phycomyces*.[17] We found that the growth rates of "stiff" mutants — defective in one of four different genes associated with growth regulation — exhibit a dominant 10.4-minute oscillation in growth rate. This growth-rate rhythm may underlie the growth fluctuations of the wild type (normal) sporangiophores as well, but they appear to be masked by the larger nonlinear fluctuations.

Max Delbrück and Phycomyces

Following Castle's early studies, Max Delbrück selected *Phycomyces* as a model organism for his investigations of sensory physiology at California Institute of Technology. Delbrück made groundbreaking contributions to biology in the field of bacterial genetics during the 1930s and 1940s and was awarded the 1969 Nobel Prize for physiology or medicine for this work.[18] In abandoning molecular biology, Delbrück bucked the trend, for the 1950s were heady days in molecular biology. In 1953, just as Delbrück was beginning his studies of the *Phycomyces* sensory system, James Watson and Francis Crick published their landmark paper that revealed the double-helix structure of DNA.[19] In fact, shortly after Watson and Crick published their famous paper, Watson moved from Cambridge University to Caltech, where he worked on the structure of tobacco mosaic virus. Although most of Delbrück's attention was devoted to *Phycomyces* at that time, a former student from Delbrück's lab relates that he continued to act as a "father confessor and Dutch uncle" of the Caltech phage group.[20]

Delbrück's first *Phycomyces* publication ("System analysis for the light growth reactions in *Phycomyces*") resulted from a collaboration with the German physicist Werner Reichardt.[21] This paper examined sensory adaptation in *Phycomyces* and established a framework for Delbrück's *Phycomyces* studies of the next thirty years. As in the visual system of humans, *Phycomyces* exhibits "light adaptation," a decrease in sensitivity that occurs upon transfer from an environment with dim light to one with bright light, and "dark adaptation," an increase in sensitivity that occurs upon transfer from an environment with bright light to one with dim light.[22]

Adaptation allows *Phycomyces* to adjust its photosensitivity to an amazing ten billion-fold change in ambient light level, from about 10^{-9} to 10 watts per square meter (Wm^{-2}). Moreover, a fully dark-adapted sporangiophore is so photosensitive that its threshold for phototropism is about 100,000-fold lower than the light level during a cloudy night during a new moon. The threshold level for phototropism in *Phycomyces* is about 100 to 1,000-fold lower than that for the phototropism of higher plants (see sidebar, "Threshold for Phototropism in *Phycomyces* and Plants"). The photosensitivity and range of adaptation of the *Phycomyces* sporangiophore are similar to those of the human eye (see table 1, p. 6).

These and other similarities of the *Phycomyces* sporangiophore and the human eye led Delbrück to hypothesize that they use a similar photoreceptive pigment.[23] Retinal, a derivative of carotene, was well known as the photoreceptive pigment in human eyes (see figure A1, Appendix), so an early hypothesis was that beta-carotene, which is abundant in the sporangiophore, was the photoreceptive pigment in *Phycomyces*. This early hypothesis was in keeping with an important point Delbrück made in a 1976 Copenhagen lecture: plants, animals, fungi, and bacteria use a very small number of molecules as photoreceptive pigments.[24]

In the late 1960s Delbrück's lab began a concerted effort to conclusively identify the photoreceptive pigment in *Phycomyces*. Initially, they isolated mutants unable to synthesize beta-

Threshold for Phototropism in Phycomyces *and Plants*

Phycomyces sporangiophores, oat seedlings, and corn seedlings are often used as model organisms in phototropism experiments.[†] The phototropic thresholds (level of light that causes a just-detectable response) of these species indicate that the *Phycomyces* sporangiophore is one hundred to one thousand times as photosensitive as oat or corn seedlings.

Species	*Illumination Protocol*	*Threshold*[‡]
Phycomyces (stage I)	10 sec, blue light (450 nm)	$\sim 10^{-6}$ Joules m^{-2}
Phycomyces (stage IV-b)	30 sec, blue light (450 nm)	$\sim 10^{-5}$ Joules m^{-2}
Avena coleoptile (oat seedling)	30 sec, blue light (458 nm)	$\sim 10^{-3}$ Joules m^{-2}
Zea coleoptile (corn seedling)	<5 min, white light	$\sim 10^{-3}$ Joules m^{-2}

carotene, the principal carotenoid in the sporangiophore. When they examined phototropism in these mutants, they found that their photosensitivity was the same as the carotenoid-containing wild type.[25] This ruled out beta-carotene as a photoreceptive pigment. Because flavin (vitamin B-2) absorbs the same wavelengths of light that are effective in phototropism and, like beta-carotene, is abundant in the *Phycomyces* sporangiophore, a flavin seemed likely to be the photoreceptive pigment.

In Delbrück's last major *Phycomyces* publication, he and collaborators demonstrated that a flavin is indeed the photoreceptive pigment that controls phototropism.[26] This 1981 publication, "Replacement of riboflavin by an analogue in the blue light photoreceptor of *Phycomyces*," resulted from a collaboration of American, German, and Japanese scientists and represents the culmination of many years of research. It is a classic publication because of its synthesis of physiology, genetics, biochemistry, and physics.

In this work, Delbrück's lab isolated a mutant that could not synthesize riboflavin, termed a "riboflavin auxotroph." They reasoned that they could feed these riboflavin auxotrophs with a riboflavin analog — roseoflavin — that is structurally similar to riboflavin but absorbs yellow light more strongly and blue and ultraviolet light less strongly than riboflavin (see figure A9, Appendix). They found that sporangiophores grown on a mixture of riboflavin and roseoflavin were more sensitive to yellow light and less sensitive to blue and UV light than sporangiophores grown on riboflavin alone. This is just what would be expected if the structurally similar roseoflavin substituted for the naturally occurring riboflavin as a photoreceptive pigment.

A Multiplicity of Photoreceptive Pigments

Just as my colleagues and I reinvestigated Castle's old observation of sporangiophore growth fluctuations in a constant environment, several labs have reinvestigated the nature of the photoreceptive pigment. It now appears that a flavin is only one of several photoreceptive pigments in the *Phycomyces* sporangiophore.

Shortly after Delbrück's death, several papers were published that showed an unusual response to red light that could not be attributed to riboflavin, because riboflavin does not have significant absorption in this region of the spectrum.[27] Subsequent studies of dark adaptation, the response initially investigated by Delbrück, uncovered an additional light effect that could not be attributed to flavin. In fact, the results of these adaptation studies seemed paradoxical, because they showed that light below the threshold for phototropism, which the investigators call "subliminal light," actually accelerates dark adaptation.[28] In other words, exposure to light accelerates adaptation to darkness.

Paul Galland and colleagues later demonstrated through a series of intricately designed experiments that red and yellow light are most effective in accelerating dark adaptation.[29] This was surprising for at least two reasons. First, this effect is not consistent with flavin as a photoreceptive pigment, because flavin absorbs red and yellow light weakly. Second, virtually all known photosensory responses in fungi are controlled by blue and UV radiation.[30] In short, these adaptation experiments indicated that the sporangiophore has a novel photoreceptive pigment that *Phycomyces* researchers apparently had not noticed in more than one hundred years of study.

Another remarkable feature of the sporangiophore's response to red and yellow light is its extraordinary photosensitivity. Whereas very dim blue light of about 10^{-9} W m^{-2} (roughly the same as the threshold for vision in humans) can elicit phototropism, the adaptation acceleration response can be elicited by light of less than 10^{-11} W m^{-2}, more than one hundred times lower.[31]

The high sensitivity to red and yellow light indicates two important features of this new photoreceptive pigment.[32] First, it must occur at a fairly high concentration in the sporangiophore, or there would not be enough pigment to absorb such low levels of light. Second, the transduction reactions that follow from absorption of light to acceleration of adaptation must occur with high efficiency. Because this new pigment exists in high concentration, it should be possible to purify it by use of a spectroscopic methodology. Such experiments seem only to be awaiting an eager and able pair of hands.

What else lies ahead for *Phycomyces* research? One researcher who performed some of the initial red light experiments told me that he gave up on *Phycomyces* after several years because he realized that its photosensory system was much more complex than he originally imagined. He noted that the sporangiophore has a photosensory system with multiple photoreceptive pigments, as well as an array of additional sensory systems for responding to gravity, wind, odors, and the presence of nearby objects. He argued that it would be easier and more fruitful to study biological responses to light in cells specialized for responding to light, such as the rod cells of the human retina.

I certainly agree that the sporangiophore, a little hair-sized organ, is not a simple biological system. But I don't agree with the rest of his argument. *Phycomyces* was the first organism in which a blue light–photoreceptive pigment was conclusively identified as a flavin, and its responses to red light suggest that it has much more to tell us about photoreception and the feedback pathways that regulate adaptation. Perhaps we need only ask *Phycomyces* the right questions.

Moreover, just as a mother does not easily abandon a child born of her flesh, I cannot easily abandon *Phycomyces*. I have often returned to that special spot in the park near my home where I sowed my *Phycomyces* spores. I have not seen any since that first year but remain hopeful that they will spring up once again, for spores can live a long time.

Suggested Reading

Cerda'-Olmedo, E., and E. D. Lipson (1987) *Phycomyces*. Cold Spring Harbor, N.Y., Cold Spring Harbor Laboratory.

Corrochano, L. M. (1999) The *Phycomyces* Web Site (www.es.embnet. org/~genus/phycomyces.html).

Kaplan, D. R., and T. J. Cook (1996) The genius of Wilhelm Hofmeister. *American Journal of Botany* 83, 1647–60.

10 Dictyostelium, *the Amoeba and the Slug*

We have a strong predilection for thinking
of organisms as adults rather than as
life cycles.
— *John Tyler Bonner*

In 1869, just as Mad King Ludwig began building his fantasy castles in Bavaria, a more down-to-earth German, a biologist named Oskar Brefeld, described his discovery of an unusual organism that he found growing on horse excrement.[1] Brefeld did not realize the significance of this new organism until he examined its unusual life cycle in the laboratory, growing it on a medium of cooked horse dung. This new organism, which Brefeld named *Dictyostelium mucoroides*, was the first species to be identified in an entire taxonomic class now called the Dictyostelia — the cellular slime molds. The cellular slime molds are so unusual that over the years there has been controversy about whether they should be classified as plants, animals, or fungi.[2] They have some features of plants, in that they make reproductive spores on erect stalks; some features of animals, in that they undergo metamorphosis and eat bacteria by phagocytosis (engulfment); and some features of fungi, in that they live on decaying plant debris. Traditionally, the cellular slime molds have been discussed in mycology textbooks, although many taxonomists consider them distantly related to the fungi.[3]

For many years, biologists must have considered Brefeld's cellular slime mold an oddity; only about twenty-five papers were published about these organisms in the following sixty-three years.[4] In 1935 Kenneth Raper published a paper describing a new species of cellular slime mold, *Dictyostelium discoideum*, that he isolated from leaf litter collected while on a camping trip in the Smoky Mountains of North Carolina.[5] When Raper studied this new species in the laboratory, he showed that its fascinating life cycle made it an almost ideal model organism for studies of cellular and developmental biology. In particular, its development resembled the metamorphosis of an animal embryo, in that simple unspecialized cells interact and change into specialized cells in a complex multicellular organism (see figure 7). As noted by John Tyler Bonner, the unique life cycle of *Dictyostelium discoideum* serves as a reminder that an organism is not merely an adult, but exists as several different stages in a life cycle.[6]

Raper's important research stimulated interest in the cellular slime molds. From 1932 to 1959, when John Tyler Bonner published a monograph on the cellular slime molds, about one hundred papers had been published on this group of organisms.[7] Since 1959, research on the cellular slime molds has increased dramatically. A recent compilation lists more than six thousand research articles on *Dictyostelium*.[8] Remarkably, as with *Phycomyces* (see Chapter 9), few of the biologists who currently study *Dictyostelium* in the laboratory have ever seen it in nature.

Modern biologists have found *Dictyostelium discoideum* to be an excellent model organism for the study of development, biochemistry, intercellular communication, and molecular genetics. The proteins and chemical pathways that control development and cell movement in *Dictyostelium* are similar to the molecules that humans and other organisms use for muscle movement and chemical signaling.[9]

Dictyostelium is also fodder for photobiologists, because several different stages in its life cycle — the amoeba and the slug —

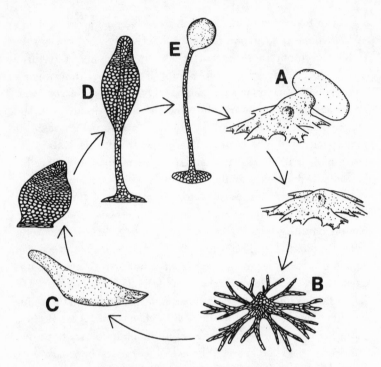

Figure 7. *Dictyostelium discoideum* life cycle.
A: spores give rise to amoebas that feed on bacteria and produce daughter cells. B: when the food supply is depleted, amoebas develop photosensitivity and then secrete and are attracted to cyclic AMP (see figure A10, Appendix), forming two-dimensional aggregation streams. C: aggregation continues until the formation of a photosensitive slug, whose movement is also controlled by cyclic AMP. D: the culmination stage occurs as cells at the slug tip push downward to form a stalk, while posterior cells migrate up the stalk to form the spore cells on top. E: eventually, a mature spore-bearing sorocarp forms and releases its spores. Drawing by Nick Colas, Syracuse, N.Y.

are responsive to light.[10] The light responses of *Dictyostelium* may also be relevant to humans, because the photoreceptive pigments appear to be porphyrins or porphyrinogens. These compounds are intermediates in the synthesis of heme, an essential component of hemoglobin and myoglobin, important oxygen-carrying proteins in human blood and muscles. Porphyrins and porphyrinogens also accumulate in patients with a metabolic disease called porphyria, causing skin photosensitivity and assorted other symptoms (see Chapter 6).

Amoeba

As with humans, *Dictyostelium* begins its life as a single cell. In its natural environment (leaf litter on the forest floor) or in a laboratory petri dish, *Dictyostelium* amoebas feed on bacteria by phagocytosis, a process in which cellular protrusions of the amoeba surround bacteria, completely engulf them, and finally digest them intracellularly. The *Dictyostelium* amoebas, apparently sensing the folic acid that is excreted by distant bacteria, migrate toward this compound, a process known as positive chemotaxis.[11] At this early developmental stage, the amoebas have a solitary existence and periodically divide by cellular fission, with the daughter cells seeking their own bacteria as food.

When a group of amoebas have consumed all the bacteria in their environment, or are transferred to a bacteria-free petri dish in the laboratory, fascinating changes occur. During the first two hours or so of starvation, the amoebas exhibit phototaxis. They move toward a weak light source (positive phototaxis) and away from a strong light source (negative phototaxis). The threshold for positive phototaxis to white light is about 10^{-4} W m^{-2}, roughly equivalent to the light level on a cloudy midnight during new moon (see table 1, p. 6).[12] The threshold for negative phototaxis to white light is about 0.1 W m^{-2}, roughly equivalent to the light level during late twilight on a clear night. The amoebas do not exhibit any orientation to light when abundant bacteria are present or when they are actively dividing.[13]

In order to better understand the light responses of amoebas, Donat Häder and colleagues focused microbeams of light on different parts of individual cells.[14] When a portion of an amoeba is irradiated with a dim microbeam of light, cellular projections called pseudopodia — literally "false feet" in Greek — develop and grow larger at the site of irradiation. In contrast, if a bright microbeam of light is aimed at an individual pseudopod, it is withdrawn from the light. Both of these responses are elicited by irradiating the cell periphery. These results suggest that a photoreceptive pigment is in the cell's plasma membrane and that amoebas turn toward light by pseudopod extension.

Häder and colleagues also determined an action spectrum for amoeba phototaxis, a graph that shows the effectiveness of different wavelengths of light and typically resembles the absorption spectrum of the photoreceptive pigment.[15] The action spectrum for amoeba phototaxis has a maximum near 400–410 nm (violet light) and several secondary peaks between 500 and 630 nm (green-red light), depending on the particular genetic strain.[16] A membrane-bound protein-pigment complex with an absorption spectrum similar to the action spectrum has been purified. The pigment has been tentatively identified as protoporphyrin IX, the same compound that accumulates in people with a metabolic disease called erythropoietic porphyria and makes their skin sensitive to light (see Chapter 6 and figure A5, Appendix).[17] Protoporphyrin IX also serves as a photoreceptor in *Escherichia coli*, a common bacterium that lives in the human intestine (see sidebar, "Porphyrin Photoreceptor in *Escherichia coli*").

Although the mechanism of the light response in amoebas is not well known, it seems possible that light absorption by protoporphyrin IX generates reactive forms of oxygen and that these toxic compounds somehow mediate the physiological response.[18] A similar chemical mechanism has been implicated in causing

Porphyrin Photoreceptor in Escherichia coli

Escherichia coli, the bacteria that naturally live in the human gut, normally "run" (move forward smoothly) when their tail-like flagella turn counterclockwise and "tumble" (move erratically) when the flagella turn clockwise. These running and tumbling motions typically alternate, so the bacteria execute a three-dimensional random walk.° In general, *E. coli* tumble more when a repellent is present and run more when the repellent is removed. A specific *E. coli* mutant, with a genetic defect analogous to that which causes erythropoietic porphyria (see Chapter 6), accumulates protoporphyrin IX† (see figure A5, Appendix). These mutants tumble when exposed to violet-blue light and run when the light is removed. This response only occurs in the presence of oxygen. This behavioral effect is likely due to light absorption by protoporphyrin IX (which strongly absorbs violet-blue light) and subsequent generation of reactive forms of oxygen that act as repellents. Wild-type (normal) *E. coli* cells also respond to blue light, but only at much higher light levels.‡

skin photosensitivity in people with various forms of porphyria.[19] Reactive forms of oxygen are often portrayed as having a negative role, as in porphyria, but there is evidence that they act as cellular messengers in many plant and animal species.[20]

After the amoebas have been starved for about two hours, they lose their response to light, and their behavior becomes increasingly dominated by cAMP (cyclic adenosine monophosphate), a compound derived from ATP (adenosine triphosphate), the energy "currency" of cells (see figure A10, Appendix,

Cyclic AMP

Cyclic AMP was discovered in 1957 by Earl Sutherland.§ This compound is derived from ATP (see figure A10, Appendix), which provides energy for many biochemical and cellular processes within the cells of all organisms. Sutherland was awarded the 1971 Nobel Prize in physiology or medicine for showing that cAMP is an important messenger for intracellular communication in many organisms. In humans, cAMP activates breakdown of glycogen and lipids and amplifies the signals from many hormones and neurotransmitters. As far as we know, the cellular slime molds are the only organisms in which cAMP plays a role in intercellular communication.

and sidebar, "Cyclic AMP"). In a group of starving amoebas, a centrally located cell secretes pulses of cAMP. This compound binds to receptors on the surface of nearby cells and causes them to move toward the focal cell for a minute or two.[21]

After a group of cells has migrated toward the focal cell, they begin to secrete their own cAMP. This, in turn, causes more distant amoebas to move toward the center of the aggregating mass of cells. These newly recruited amoebas then secrete cAMP, and the process continues. A migrating cell typically secretes cAMP for several minutes, and the mass of migrating cells often form beautiful patterns of concentric waves, with each wave consisting of hundreds or thousands of cells. The amoebas do not respond to their own cAMP because they are fully adapted to this chemotactic signal by the time they secrete it themselves.[22] The phosphodiesterase that the cells secrete also plays an essential role in wave propagation.[23]

Slug

Dictyostelium amoebas continue to aggregate until a large mass of many thousands of undifferentiated cells — the pseudoplasmodium — forms (see figure 7).[24] Most biologists simply call this stage the "slug," because it resembles the familiar garden slug and even leaves a slime trail as it travels along. (Garden slugs are gastropod mollusks, close relatives of snails, unrelated to the cellular slime molds.) The *Dictyostelium* slug is typically about a millimeter in length, barely visible to the naked eye, and moves at about the same speed as individual amoebas, 0.2 to 2.0 millimeters per hour.

The *Dictyostelium* slug, like the *Phycomyces* sporangiophore, responds to a variety of environmental signals, including light (phototaxis), temperature (thermotaxis), wind (rheotaxis), and various chemicals (chemotaxis), including ammonia (NH_3) and an as yet unidentified "slug turning factor."[25] In nature all of these responses presumably enable the slug to move toward the soil surface, where the subsequent life-cycle stage, the spore-bearing sorocarp, can more successfully disperse its spores.

The tip of the slug functions like a brain, in that it coordinates the continuous movement of the thousands of constituent slug cells. As with amoebas, these movements are governed by cAMP. In the case of the slug, the cell movements are called "scroll waves," the three-dimensional generalization of the two-dimensional spiral waves seen in aggregating amoebas.[26] Cyclic AMP, ammonia, and the slug turning factor, all of which are produced by the slug itself, apparently control phototaxis by acting as chemical messengers that facilitate communication among the thousands of slug cells.[27] Moreover, actin and myosin, the same proteins that power movement of our own muscles, play important roles in controlling slug movement.

Just how does the slug sense light? When the slug is irradiated with a horizontal beam of light, a focal spot forms on its distant side, much in the way the human lens focuses light onto the retina.[28] Slugs exhibit negative phototaxis to UV radiation (260–90 nm), presumably because this radiation is absorbed by substances within the individual cells of the slug, preventing formation of a focal spot.[29] Paul Fisher and colleagues suggested that light induces synthesis of a slug turning factor, a low molecular weight compound that repels slugs.[30] In other words, according to Fisher, slugs move toward the light by moving away from the slug turning factor that is synthesized at the focal spot.

The slug is extremely sensitive to light, nearly as sensitive as our own eyes. The threshold for slug phototaxis is about 10^{-6} $W\,m^{-2}$ of white light, roughly 100 times lower than the threshold for positive phototaxis in amoebas and also about 100 times lower than the light level on a cloudy midnight during new moon.[31] The action spectrum for slug phototaxis is somewhat different from that for amoeba phototaxis, suggesting that amoebas and slugs use different pigments to control phototaxis.[32] In support of this, a *Dictyostelium* mutant has been generated that exhibits normal phototaxis as an amoeba but does not exhibit any phototaxis as a slug.[33]

The identity of the photoreceptive pigment for slug phototaxis is still uncertain, although a high–molecular weight heme protein or a protein that contains a flavin (vitamin B-2; see figure A9, Appendix) and a cytochrome (porphyrin-based compound; see figure A5, Appendix) have been proposed.[34] Warren Butler, one of the team of scientists who first purified phytochrome, an important photoreceptive pigment in plants (see Chapter 7), worked on the preliminary purification of the slug photoreceptor more than twenty-five years ago.[35] Since then, the identity of the photoreceptive pigment has not been confirmed by molecular genetics studies, in which disruption of the specific gene that codes for the photoreceptive molecule has been shown to alter the physiological response to light.

Many studies, however, have employed genetic mutants that exhibit altered phototactic behavior. Three methods have been used to estimate the number of genes specifically involved in slug phototaxis. Each gives an estimate of about twenty phototaxis *(pho)* genes:

- A statistical method known as the "maximum likelihood method" estimates that about seventeen genes are specifically important for slug phototaxis.[36]
- The number of phototaxis mutants isolated by chemical mutagenesis is about one in four hundred, or about 25-fold higher than expected if a single gene were involved.[37]
- The number of mutants isolated by plasmid-mediated mutagenesis (technically, nontargeted disruption with integrating plasmid DNA) is about one in six hundred, or about 20-fold higher than expected if a single gene were involved.[38]

Sorocarp

Under the proper environmental conditions, the slug develops into a sorocarp, a "fruiting body" that typically has a small base, an erect stalk that is one to two millimeters long, and a ball-like cap that is less than half a millimeter in diameter but is filled with many thousands of spores (see figure 7). The exact dimensions of the sorocarp depend on the environmental conditions under which it develops. For example, Kenneth Raper, the discoverer of *Dictyostelium discoideum,* reported more than fifty years ago that light of about 0.1 W m^{-2}, roughly equivalent to the level during late twilight on a clear night, causes *Dictyostelium* to develop smaller aggregations and smaller sorocarps.[39]

The developmental fate of the 100,000 or so cells in the developing sorocarp is determined at the slug stage. Cells in the slug tip (~20 percent) are fated to become stalk cells, while base cells (~1–2 percent) and the remaining slug cells (~80 percent) are fated to become spore cells. All of the 20,000 or so cells that make up the stalk eventually die, while the genetically identical spore cells develop inside the cap.[40] John Tyler Bonner, a veteran researcher of *Dictyostelium,* has argued that the sacrifice of so many cells indicates that the sorocarp has great adaptive value.[41] Because we know so little of the natural history of the cellular slime molds, it is difficult to know exactly why the sorocarp has such great adaptive value. We do know, however, that sorocarps form on the soil surface, so it seems likely that worms, insects, and other arthropods brush against sorocarps and then inadvertently disperse the microscopic slime mold spores in their daily travels. In support of Bonner's hypothesis of the adaptive value of the sorocarp, there is incredible diversity in sorocarp morphology among the different species of *Dictyostelium.*[42] Such diversity is an expected result of intense pressure by natural selection.

The *Dictyostelium* sorocarp resembles the *Phycomyces* sporangiophore, even though these organisms are only distantly related.[43] Because both the amoeba and the slug are very sensitive to light and exhibit well-coordinated responses to light, it might be expected that the sorocarp, like the *Phycomyces* sporangiophore and the stems of young seedlings, would also bend toward the light as some sort of adaptation for spore release. In fact, the rising sorocarp does lean toward the light, though little research has been devoted to this response because slug orientation is so much easier to measure in the laboratory.[44] After the *Dictyostelium* sorocarp has released its spores toward the light, these spores eventually germinate and begin life anew as amoebas. In fact, every hour of every day in forest soils throughout the world, the different species of *Dictyostelium* pass through their life cycles unnoticed, with amoebas and slugs migrating in response to chemicals, temperature, wind, and light.

Suggested Reading

Bonner, J. T. (1994) The migration stage of *Dictyostelium discoideum:* behavior without muscles or nerves. *FEMS Microbiology Letters* 120, 1–8.

Dictyostelium WWW Server (1997) (dicty.cmb.nwu.edu/dicty/dicty. html).

Fisher, P. R. (1997) Genetics of phototaxis in a model eukaryote, *Dictyostelium discoideum. BioEssays* 19, 397–407.

Kessin, R. H., et al. (1992) The development of a social amoeba. *American Scientist* 80, 556–65.

Pattern formation in *Dictyostelium discoideum* (1998) (www.zi.biologie .uni-muenchen.de/zoologie/dicty/dicty.html). Includes MPEG videos of amoeba aggregation and slug phototaxis.

Raper, K. B. (1984) *The Dictyostelids.* Princeton, Princeton University Press.

11 High Hopes for Hypericin

I got high hopes!
— *Frank Sinatra*

Saint John's wort has been in the news quite a lot in recent years. In the late 1980s, several studies showed that red pigments derived from this herb — hypericin and pseudohypericin — inactivated or blocked the reproduction of certain retroviruses in mice. This led to speculation that these compounds might be effective against human retroviruses, including the human immunodeficiency virus (HIV), the virus associated with AIDS.[1] Indeed, subsequent preclinical studies showed that hypericin could destroy the equine infectious anemia virus and the related human immunodeficiency virus.[2] By the start of 2000, no clinical studies had demonstrated that hypericin or Saint John's wort is an effective treatment for AIDS, but research in this area continues.[3]

Another line of research has shown that Saint John's wort may effectively treat mild to moderate depression and seasonal affective disorder (see Chapter 5) with fewer side effects than certain standard antidepressant drugs.[4] Many studies have examined the biochemical basis of this antidepressant effect. Surprisingly, crude extracts of the plant exhibit only weak activity in biochemical assays related to the neurochemical mechanisms of widely used antidepressants, such as inhibitors of monoamine oxidase (nardil and parnate) and serotonin reuptake (prozac, zoloft, and paxil).[5] Crude extracts of Saint John's wort have shown affinity for nerve cell receptors that normally bind to gamma-aminobutyric acid (GABA), an important neurotransmitter in the brain.[6] Barbiturates, benzodiazepines (Valium and Xanax), and alcohol all exert their profound effects on the central nervous system by binding to GABA receptors.[7] At present, it must be concluded that the mechanism by which Saint John's wort alleviates depression is unknown. It is possible that the combined action of numerous mechanisms accounts for the overall effect. Future laboratory studies must be directed toward determination of the active ingredient(s) and its mechanism of action, and clinical studies should employ more standardized doses, longer-lasting trials, and comparisons with more modern antidepressants.[8]

Study of the myriad of possible therapeutic effects of Saint John's wort and hypericin continues, as it has for centuries (see sidebar, "Botany and Early Uses of Saint John's Wort"), and gives many researchers high hopes for Saint John's wort.[9] At the same time, photobiologists have been studying the function of naturally occurring hypericin-like pigments in two free-living microorganisms, *Stentor coeruleus* and *Blepharisma japonicum* (see figure 8). The hypericin-like pigments in these microorganisms, known as stentorin and blepharismin, are photoreceptive pigments that control cellular locomotion (see figure A11, Appendix). Understanding the mechanism by which these pigments control cellular movements may clarify the mechanisms by which hypericins act as antiviral agents and Saint John's wort as an antidepressant. At a more fundamental level, study of the light-induced movement responses of *Stentor* and *Blepharisma* may provide a more basic understanding of the chemistry of hypericins, a unique class of biological pigments.

Biology of the Ciliates

Stentor coeruleus and *Blepharisma japonicum* are classified as Ciliates, a taxonomic group of single-celled, nonphotosynthetic organisms that are covered with cilia, short whiplike appendages

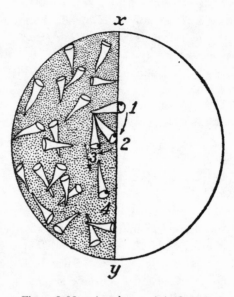

Figure 8. Negative phototaxis in *Stentor.*
Herbert S. Jennings, an early researcher of behavior in ciliates and
other single-celled organisms, made this line drawing of the view through
a microscope of *Stentor coeruleus.* The illustration shows that cells move
about randomly in darkness, but when the anterior end of a cell enters
a lighted region (1), it turns away (2) and swims back toward darkness
(3 and 4). The lighted area remains empty because of this negative
phototactic response.

that beat back and forth to propel them through the water or
sweep food into their mouthlike openings. The cilia are typically
arranged in precise and species-specific patterns on the surface
of the cell and are firmly embedded in a one-micrometer-thick
protein-rich layer that coats the outside of the cell membrane.
There are about eight thousand known species of ciliates and
probably many more unidentified species. The ciliates have little
economic significance, with the possible exception of *Balantidium*

Botany and Early Uses of Saint John's Wort

Saint John's wort *(Hypericum perforatum)* is a perennial herb that grows to a foot or more in height and during July or August produces flowers, each with five bright yellow petals.° It is native to Europe but has become naturalized throughout most of the United States and typically grows along roadsides and in meadows, rangelands, and pastures.† It is commonly considered a noxious weed because the hypericin in its flowers and leaves can cause "hypericism" — inflammation following exposure to sunlight — in the grazing animals that eat these plants.‡ Hypericism is caused by the formation of reactive oxygen species following light absorption by hypericin.

Many ancient herbal doctors, such as Hippocrates (460–377 B.C.), Dioscorides (A.D. 41–68), and Galen (A.D. 150–200), have recommended the herb for the treatment of numerous ailments.§ Many of these so-called treatments were formulated according to the now discredited "doctrine of signatures," which said that a plant's outward appearance indicates its therapeutic value. One such treatment employing Saint John's wort is given by John Gerard, a well known sixteenth-century century English herbalist. Gerard wrote about an extract of Saint John's wort that presumably contained the red pigment hypericin as well as other compounds:

> *The leaves, flowers, and seeds stamped, and put into a glass with oyle olive, and set in the hot sunne for certain weeks together, and then strained from these herbes, and the like quantity of new put in, and sunned in the like manner, doth make an oyle the color of blood, which is a most precious remedy for deep wounds and those that are thorow the body, for sinews that are pricked, or any wound with a venomed weapon.*#

coli, a species that inhabits the human gut and causes a rare form of dysentery. *Stentor* and *Blepharisma* belong to a group of ciliates called the polyhymenophorans, all of which have complex ciliary structures.[10] The polyhymenophoran ciliates are an ancient group, as fossils of these ciliates have been dated to 100 million years ago.

Biologists have studied light-induced responses of the ciliates for more than one hundred years.[11] Sixteen species of ciliates exhibit some form of phototaxis, moving toward the light (positive phototaxis), away from the light (negative phototaxis), or perpendicular to the light direction (transverse phototaxis).[12] Of these sixteen species, five — including *Stentor* and *Blepharisma* — are brightly colored due to the presence of vesicles that contain pigment granules. These vesicles lie between the rows of cilia, directly beneath the cell membrane. In all five brightly colored species, the pigment granules presumably contain the photoreceptive pigment that controls phototaxis.

The *Stentor* cell is typically about 0.35 millimeters long, although some strains can grow up to 2 millimeters in length. It assumes a pearlike shape when swimming but is otherwise shaped like a trumpet.[13] This species is named after the ancient Greek warrior Stentor, who, according to Homer, had a trumpetlike voice and "could cry out in as great a voice as fifty other men."[14] *Stentor* cells have longitudinal rows of granules (0.3 to 0.7 micrometers in diameter) that are blue-green in color because they contain the pigment stentorin, which strongly absorbs red and ultraviolet radiation.[15] *Blepharisma* is about the same size as *Stentor* but is spindle-shaped.[16] *Blepharisma* is named after the Greek word for eyelid *(blepharon),* because its long cilia resemble eyelashes. These cells have pigment granules (0.35 micrometers in diameter) that make them appear red under dim light because of the fluorescence emitted by the pigment blepharismin.[17] *Blepharisma* turns blue-green under strong light, however, because blepharismin is transformed to "blue-blepharismin," which is only weakly fluorescent.

Behavioral Responses to Light

Stentor coeruleus was one of the first ciliates used to study light-induced movement responses.[18] A ring of hairlike cilia surrounds its large open end, which serves as a mouth and anus and is most sensitive to light and other stimuli. More sparsely distributed cilia cover the rest of its surface. At the narrow posterior end lies the "foot," which often attaches to stationary objects. *Stentor* avoids the light and tends to collect in shaded regions by using a trial-and-error method — it constantly rotates about its long axis to sample the light environment, then aims toward the darkness.

Stentor and *Blepharisma* exhibit step-up photophobic responses (see sidebar, "Light Control of Microorganism Movements"): an increase in the light level causes the cilia to reverse the direction of their beating, so that the cell stops, reverses direction, and moves out of the light. These two species also exhibit negative phototaxis, wherein they move away from a directional light source.[19] When cells are irradiated with a light beam that converges to a focal point and then diverges, they swim through the bright focal point and away from the light source.[20] This indicates that they truly sense the light direction and not merely the fluence rate ("intensity") of the light.[21] In nature, the step-up photophobic response and negative phototaxis presumably allow *Stentor* and *Blepharisma* cells to move toward the bottom of a pond, where they can avoid zooplankton predators or develop into cysts — dormant and resistant capsules.

Stentor and *Blepharisma* are killed when exposed to very bright light. Thus while a one-second pulse of light at 0.1 W m^{-2}, roughly equivalent to the light level during late twilight (see table 1, p. 6), elicits a stop response in *Stentor*, light of 5000 W m^{-2},

> ### *Light Control of Microorganism Movements*
>
> Light controls the movement of many microorganisms, and these movements are commonly classified as photophobic responses, photokineses, or phototaxes.** In photokinesis, cells sense the fluence rate ("intensity") of light; in the photophobic response, they sense a change in fluence rate; and in phototaxis, they sense the light direction. More formally, these terms may be defined as:
>
> - *Photokinesis,* light-induced change in cell speed or frequency of alterations in cell direction.
> - *Photophobic response,* photon fluence rate change–induced alteration of cell movement, typically a "stop" and reversal of direction.
> - *Phototaxis,* movement oriented with respect to the direction of light that is typically toward (positive), away from (negative), or perpendicular to (transverse) the light source.

roughly equivalent to the light level during a clear midday, kills the cells.[22] Under high levels of light, stentorin generates reactive oxygen species that react with proteins, lipids, and other molecules, and this damage eventually kills the cells. Light-induced generation of reactive oxygen species is also responsible for the phototoxic effects of hypericin, many of the symptoms of the metabolic disease porphyria (see Chapter 6), and the harmful effects of bright sunlight on plants (see Chapter 15).[23]

The effect of very bright light on *Blepharisma* is more complicated. Cells grown under dim light (which appear red due to fluorescence from the pigment blepharismin) are killed by light of 30 W m^{-2} because reactive oxygen species are generated. If the cells are first exposed to light of about 3 W m^{-2}, however, they are transformed into blue or colorless cells that can survive treatment with light of 1500 W m^{-2} for twenty minutes or more.[24]

Because stentorin and blepharismin, the hypericin-like pigments of *Stentor* and *Blepharisma*, are responsible for the toxic effects of bright light, one may reasonably ask: why synthesize these pigments in the first place? A possible answer is that in nature, *Stentor* and *Blepharisma* use these hypericin-like pigments to control movement of the cells out of dim light so they can avoid the phototoxic effect of very bright light.[25]

Chemistry of the Response to Light

The action spectrum (a measure of the effectiveness of different wavelengths) for the photophobic response in *Stentor coeruleus* has a major peak near 610 nm (red light), with minor peaks near 550 nm (green light) and 480 nm (blue light).[26] This is similar to the absorption spectrum of a stentorin-binding protein that has been extracted from *Stentor* cells.[27] In fact, the stentorin-binding protein is part of a large assembly of proteins that includes other proteins that do not bind stentorin.[28] Blepharismin has been less intensively studied than stentorin. Research has shown, however, that the action spectrum for the photophobic response in *Blepharisma*, which has a major peak near 590 nm and minor peaks near 540 nm and 480 nm, is similar to the absorption spectrum of a blepharismin-binding protein.[29]

As with all other photoreceptive pigments, the absorption of light by stentorin and blepharismin must generate chemical signals that are eventually converted into a physiological response — in this case, cell movement. While the generation of reactive forms of oxygen is clearly responsible for cell death under high levels of light, this reaction does not appear to play a role in the movement responses.[30] Instead, an increase in acidity (pro-

ton concentration) following absorption of light by these pigments appears to be a crucial chemical reaction needed for cell
movement.[31] Drugs that alter the internal acidity of the cells
also reduce their responsiveness to light but have little effect on
the cells' overall viability or responsiveness to other stimuli.[32]
The light-generated increase in acidity in *Stentor* apparently
arises from the rapid ($\sim 10^{-12}$ second) transfer of a proton from
the pigment stentorin to the associated stentorin-binding protein, and the subsequent release of this proton to the cytoplasm.[33]

Recent studies suggest that the increase in acidity that follows light absorption also alters the electrical potential across the
cell membranes of *Stentor* and *Blepharisma*. In particular, when
microelectrodes (finely tapered glass tubes filled with a solution
that conducts electricity) are inserted into these cells, measurements show that the membranes have a resting potential of about
-45 to -60 millivolts, meaning that the inside is more negatively
charged than the outside.[34] About two-tenths of a second after a
pulse of light, the membrane undergoes a transient depolarization by about 20 millivolts, so that the difference between the inside and outside is smaller.[35]

Such electrochemical responses are ubiquitous among organisms. Light depolarizes the photoreceptor cells of many invertebrates, even though they are only distantly related to ciliates
and use a rhodopsin as the photoreceptive pigment.[36] In contrast, the rods and cones in the eyes of humans and other vertebrates hyperpolarize (become more negative) upon illumination.
Our own nerve cells have a resting potential of about -70 millivolts and are transiently depolarized by a stimulus to about $+30$
millivolts before returning to their resting state.

A recent hypothesis links the acidifying effect of stentorin
and blepharismin, the depolarization of the membrane, and cell
movement.[37] According to this hypothesis, an increase in cellular
acidity alters certain membrane-bound proteins that regulate the

flow of ions into and out of the cell, and this depolarizes the cell membrane. Membrane depolarization, in turn, alters the properties of proteins responsible for transport of calcium ions into the cell, so that the cells take up more calcium. The increased concentration of intracellular calcium then alters certain biochemical reactions that control movement of the cilia. Experiments with specific inhibitors suggest that cGMP (cyclic guanosine monophosphate) plays an important role in transforming the light signal into a biochemical signal that controls cilia movement.[38] This same compound is also important in controlling the vision of vertebrates and invertebrates.[39] A related compound, cAMP (cyclic adenosine monophosphate), is important in controlling movement of the slime mold *Dictyostelium* (see Chapter 10).

Some people may think that study of the light-induced movement responses of insignificant microorganisms like *Stentor* and *Blepharisma* is an arcane pursuit with little significance to problems in the real world. But the photoreceptive pigments in these organisms resemble hypericin, which has shown promise as an antiviral agent. Moreover, hypericin is also present in Saint John's wort, which appears to be an effective antidepressant. In addition, the chemical pathways that *Stentor* and *Blepharisma* use to convert the light signals received by their pigments into physiological movement responses share many features with the physiological pathways used for vision, hormone signaling, and other responses in humans.

Suggested Reading

Giese, A. C. (1973) *Blepharisma: The Biology of a Light-Sensitive Protozoan.* Stanford, Stanford University Press.

Jennings, H. S. (1906) *Behavior of Lower Organisms.* New York, Columbia University Press.

Kuhlmann, H.-W. (1998) Photomovements in ciliated protozoa. *Naturwissenschaften* 85, 143–54.

Mast, S. O. (1911) *Light and the Behavior of Organisms*. New York, Wiley.

Rosenthal, N., M. Nordfors, and P. McWilliams (1998) *Saint John's Wort: The Herbal Way to Feeling Good*. New York, Harpercollins.

Tartar, V. (1961) *Biology of Stentor*. New York, Pergamon.

12 Turning on a Butterfly

And wisdom is a butterfly
And not a gloomy bird of prey.
— *W. B. Yeats*

We are all used to thinking that animals see with their eyes, but some animals have photoreceptors in rather unusual and unexpected places.[1] The sea star, though without a head or a brain, has photoreceptive eyespots near the end of its arms.[2] Birds and lizards, whose thin skulls let in light, have photoreceptor cells in the pineal glands of their brains, cells that resemble the highly specialized cone cells found in the retinas of vertebrates' eyes.[3] Many roundworms have specialized light-sensitive epidermal cells, as well as eyes with lenses and retinas.[4] But the animal with photoreceptors in the most unusual of all places is *Papilio xuthus*, the Japanese yellow swallowtail butterfly. These butterflies have photoreceptors on their genitalia.

To fully appreciate the function of these genitalic photoreceptors, one first needs to understand something about animal genitalia and their evolutionary significance.

Animal Genitalia

The male genitalia of butterflies and many other animals tend to be very complex in form and to vary greatly among closely related species.[5] In fact, the penis of some animal species is the most morphologically complex organ of the body. In contrast, female genitalia tend to be simple in form and to vary little among closely related species. This is why animal taxonomists, particularly arthropod and insect taxonomists, often use male genitalia to identify and classify closely related species.[6]

Complex male genitalia are found only in species that engage in internal fertilization — that is, those in which the male has an intromittent organ.[7] An interesting exception is the seahorse, the female of which has the intromittent organ. Complex male genitalia are not found in species that engage in external fertilization, such as echinoderms (sea stars, sea urchins, and related species) and most species of fish. In the few species of fish that do engage in internal fertilization, the males have penises with complex morphology.

The complex morphology of male genitalia and the significant differences in closely related species indicate that male genitalia have evolved rapidly and divergently.[8] Moreover, that this occurs throughout the animal kingdom suggests a general cause rather than a series of separate and specific causes.

William G. Eberhard has devoted considerable attention to the rapid and divergent evolution of animal genitalia. His book, *Sexual Selection and Animal Genitalia,* whose text and pictures will fascinate biologists and nonbiologists alike, brings together the many diverse but mostly descriptive studies of animal genitalia by placing them into a unifying evolutionary context.[9]

Sexual Selection

In his book, Eberhard argues that evolution by sexual selection is responsible for complex male genitalia.[10] Sexual selection is the competition for access to mates among members of the same gender and typically leads to the evolution of sexual dimorphism — that is, different male and female forms. More specifically, sexual selection results from competition for females' eggs and the opportunity to use them to make offspring.[11] Sexual selection was a major theme of Darwin's second great work, *Descent of Man, and Selection in Relation to Sex.*[12] Although Darwin did not apply his theory of sexual selection to animal genitalia, he did show that sexual selection, a process he considered distinct from natural selection, is responsible for the evolution of numerous other elaborate male structures.

Darwin identified two key aspects of sexual selection. The first is male-male competition, in which males compete with each other for reproductive access to females; this leads to the evolution of male "weapons," such as the antlers of the moose. The second is female choice, which leads to the evolution of showiness in males, such as the elaborate tail feathers of the peacock. Because most of the biologists who have studied sexual selection have themselves been males, there may have been an overemphasis upon the importance of male-male competition compared with female choice. In fact, Eberhard argues that female choice, not male-male competition, is responsible for the rapid and divergent evolution of male genitalia.[13]

According to the "good genes" mechanism of evolution by sexual selection, an elaborate male structure may evolve something like this: At some primeval time, males who had some genetically determined elaborate structure were more fit (perhaps better at gathering food or defending themselves) and thus had better genes than other males. Females who mated with the males that had the elaborate structure would therefore leave more offspring than those who mated with other males, and the number of males with the elaborate structure would increase. As time passed, in a process known as "runaway sexual selection," females would increasingly favor males who had ever more elaborate structures, and rapid and divergent evolution of the structure would occur. Thus, as originally argued by Darwin, sexual selection often favors a "useless" male structure, such as the extravagant tail of a peacock, even though that structure has little or no relation to the male's ability to gather food or defend itself.[14]

Evolution of Animal Genitalia by Sexual Selection

The penis, according to Eberhard, is not simply an organ designed to introduce sperm into the female but is also an "internal courtship device" that has evolved by runaway sexual selection to induce the female to use her mate's sperm to fertilize her eggs.[15] Eberhard proposes that a female will favor males whose genitalia have complex morphology because they can stimulate her better, causing her to prolong copulation, or because they trigger certain essential reproductive processes necessary for successful fertilization. Both of these responses would increase the probability of fertilization by a male with complex genitalia.

Abundant evidence supports one of these two functions for male genitalia in numerous species.[16] In many mammals, the movement of sperm into the fallopian tubes is aided by stimulating movements within the female reproductive system, such as a penis moving in and out during copulation. In rats, if the male ejaculates inside the female prior to four intromissions (in-and-out movements), the female transports less sperm to her uterus. Some male rodents have small spines on their penises whose most likely purpose is stimulation of the female during copulation. Many male free-living flatworms have rows of many penises, even though there is only a single female opening; the most likely function of all these penises is stimulation of the female.

Eberhard examined hundreds of different insect species in which specific behavior by females following copulation aids sperm movement and subsequent fertilization. Many male insects engage in complex copulatory behavior, which is best interpreted as having evolved to stimulate the female and ensure fertilization. Many male butterflies, for example, including the Japanese yellow swallowtail butterfly *(Papilio xuthus)*, move their genitalia in and out of the female during copulation, a movement that certainly stimulates the female.[17]

A traditional view in biology (favored by male biologists)

was that females of most species mate with only one male, whereas males tend to have many mates. Biologists now recognize, however, that the females of many species have multiple mates.[18] This is significant, because females must take multiple mates for female choice to have an important role in the evolution of male genitalia. In fact, an important prediction of Eberhard's theory is that the morphology of male genitalia is subject to stronger sexual selection and will have more complex morphology in species whose females take more mates. This has been demonstrated in many species. Investigation of 130 species of primates, for example, shows that elaborate penis morphology is found in species whose females have multiple mates.[19] Similarly, Eberhard has shown that among the different species in the butterfly genus *Heliconius*, species in which males have more elaborate genitalia also have females who most frequently have multiple mates.

Japanese Yellow Swallowtail Butterfly

Shortly before the publication of Eberhard's treatise on the evolution of animal genitalia, Kentaro Arikawa and his colleagues of Yokohama City University discovered photoreceptors on the genitalia of the Japanese yellow swallowtail butterfly.[20] In particular, they found that male and female butterflies have two photoreceptive spots on each side of their genitalia. Following the publication of Eberhard's book, Arikawa and colleagues have made considerable progress in their studies of the genitalic photoreceptors of these butterflies.[21]

But the genitalic photoreceptors are not the only interesting feature of light detection by this butterfly. Its eyes also have a system of color vision that, in some respects, is superior to our own. Other animals see the world and its colors differently from the way humans do (see Chapter 3). Humans have three cone pigments, and we live in a world with three primary colors (red, green, and blue). Most birds live in a world that has greater color complexity than our own, for they have an additional pigment that absorbs in the ultraviolet range.[22]

The visible world of the Japanese yellow swallowtail butterfly *(Papilio xuthus)* is even more complex. Its eyes have five pigments, so these butterflies have the potential to see five primary colors: ultraviolet (~360 nm), violet (~400 nm), blue (~440 nm), green (~520 nm), and red (~600 nm).[23] Recent behavioral tests have proven that these butterflies do indeed have color vision, though it has not yet been proven that they can actually see five primary colors.[24] One possible function of the ultraviolet and violet photoreceptors is that they enable these butterflies to see flowers that have specially designed pigmented spots (nectar guides) that reflect ultraviolet radiation and point to the nectar and pollen within.[25] Female Japanese yellow swallowtail butterflies also have a special striped pattern on their hind wings that reflects ultraviolet radiation.[26] Males may use the ultraviolet receptors in their eyes to detect this pattern, because they typically approach females from behind during courtship and mating.

Butterfly Genitalia

The eyes of the Japanese yellow swallowtail butterfly, like those of humans, function in image formation. The genitalic photoreceptors have a rather more rudimentary function: they simply detect the brightness of light. Arikawa and colleagues found that near-ultraviolet radiation (~380 nm) stimulates an electrical response in certain nerve fibers that emanate from the genital region.[27] In further studies, they probed the genital region with a 1 mm-diameter spotlight and identified two photoreceptive sites on each side of the genitalia of males and females.[28]

In males, the genitalic photoreceptors are on the scaphium (a flaplike structure on the back surface of the tuba analis) and on the diaphragm.[29] These photoreceptors are visible when the valvae (special lobes that enclose the penis) are open but not when they are closed. Light from the posterior end of the male, however, can pass through the small space between the closed valvae and stimulate his genitalic photoreceptors. In females, the genitalic photoreceptors are on the papilla analis and on a membrane between the papilla analis and the lamella postvaginalis (see figure 9).[30]

Anatomical studies by Arikawa's lab showed that the photoreceptive spots in males and females are covered by a clear cuticle (thin, outer layer of skin) that is unoccluded by hair.[31] The surrounding regions have an opaque cuticle that is occluded with hair. Dissection of the photoreceptive areas demonstrated that the cells underneath the epidermis have a layered internal membrane, a general characteristic of photoreceptor cells found in animal eyes and other organs.[32]

Light stimulation of the genitalic photoreceptors alters the activity of specific abdominal motoneurons that are connected to five or more pairs of abdominal muscles.[33] In fact, light stimulates the electrical activity of some motoneurons and inhibits the activity of others, so the behavioral response to light appears to be well-coordinated.

A film of the mating ritual of the Japanese yellow swallowtail butterfly presented at the 1994 meeting of the Zoological Society of Japan shows that after a male finds his mate, he curls his abdomen and opens his valvae, exposing the penis.[34] Then he moves into a copulatory position and grasps the female with his valvae and superuncus, a structure near the penis. After this, the male and female typically copulate, with the male thrusting his penis in and out, for about an hour.[35]

Before the male inseminates his mate, he typically opens and closes his valvae several times in succession. The purpose of this may be for stabilization of his copulatory posture or for stimulation of his mate. The photoreceptors on the male's scaphium are apparently used to monitor the copulatory posture by detect-

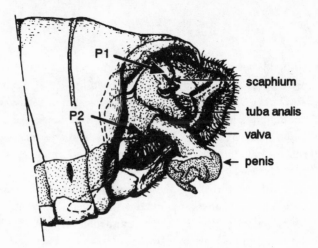

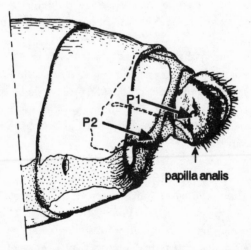

Figure 9. Genitalia and photoreceptive sites of *Papilio xuthus*.
Top: tip of the male abdomen, showing the scaphium, a
flaplike structure on the back surface of the tuba analis; the
valva, a special lobe that encloses the penis; and the
photoreceptive sites, P1 and P2. Bottom: tip of the female
abdomen, showing the papilla analis and the photoreceptive
sites, P1 and P2. Drawings courtesy of Kentaro Arikawa,
Yokohama City University.

ing whether a particular body part is covered. The electrophysio-logical response from the male's P1 genitalic photoreceptor is much lower when he is in the correct copulatory position. Appar-ently, an improper position allows light to reach the male's geni-talic photoreceptors, providing electrophysiological stimulation, and causes him to modify his position.

Light stimulation of the male's genitalia, which opens his valvae, is essential for the copulation ritual.[36] When the male's genitalic photoreceptors are occluded with black paint or ablated by a fine soldering iron, he is no longer able to grasp the female's genitalia with his valvae and does not copulate.[37] A control ex-periment with clear paint showed no effect on mating behavior. In contrast, ablation of the female's genitalic photoreceptors had no effect on copulation.

Research on genitalic photoreceptors has focused on a sin-gle species of butterfly as a model organism. Arikawa's lab has found, however, that many other species of butterflies native to Japan also have genitalic photoreceptors.[38] Thus it is likely that some of the butterflies found in our own backyards have genitalic photoreceptors.

It may seem peculiar that an animal has photoreceptors on its genitalia and that butterflies require sunlight, as well as one another, for consummation. On the other hand, we humans con-sider moonlight and a starry sky to be romantic. Perhaps this but-terfly's requirement for light is not so peculiar after all.

Suggested Reading

Arikawa, K., et al. (1980) Multiple extraocular photoreceptive areas on genitalia of the butterfly *Papilio xuthus*. *Nature* 288, 700–702.

———— (1996) Light on butterfly mating. *Nature* 382, 119.

Butterflies: Papilionidae (www.asahi-net.or.jp/~ak5t-kmn/papil/a-pa-pil.html). Includes pictures of *Papilio xuthus*.

Eberhard, W. G. (1985) *Sexual Selection and Animal Genitalia*. Cambridge, Harvard University Press.

————— (1996) *Female Control: Sexual Selection by Cryptic Female Choice.* Princeton, Princeton University Press.

Kinoshita, M., et al. (1999) Colour vision of the foraging swallowtail butterfly *Papilio xuthus. Journal of Experimental Biology* 202 (pt. 2), 95–102.

Lepidoptera List (scientific names) (www.funet.fi/pub/sci/bio/life/warp/lepidoptera-index-x.html). Includes pictures of *Papilio xuthus.*

Scriber, J. M., et al. (eds.) (1995) *Swallowtail Butterflies: Their Ecology and Evolutionary Biology.* Gainesville, Fla., Scientific Publishers.

13 Blue Moons and Red Tides

O the opal and the sapphire
Of that wandering western sea
— Thomas Hardy

In September 1883 many residents of the tropics were amazed when they looked skyward. The sun and moon were blue! These heavenly bodies returned to their original yellow-white color by October, but people the world over noted that their sunsets were particularly red and beautiful for the next year or so.[1]

In September of the following year, residents along the gulf coast of Florida were shocked when they looked out to sea. During the day, they saw that their ocean had turned red and that thousands of dead fish had washed ashore.[2]

Blue moons and red tides! What could cause this strange change of colors?

On August 27, 1883, four gigantic volcanic explosions, which were heard from more than three thousand miles away, occurred on the island of Krakatoa and expelled about five cubic miles of volcanic debris into the atmosphere.[3] As this light-scattering volcanic dust engulfed the earth, an unusual optical effect called the "blue moon phenomenon" occurred.[4] We now know that blue moons are caused by "anomalous aerosol light scattering," a type of light scattering in which light extinction increases with wavelength.[5] In other words, when the white light from the sun and moon entered the earth's atmosphere in September 1883, airborne volcanic dust from Krakatoa scattered away the long wavelength light (red) but had little effect on the short wave-length light (blue), making the sun and moon appear blue. The anomalous aerosol light scattering was so strong in September 1883 that it overwhelmed the normally dominant molecular scat-tering, which is much stronger for blue light (it is proportional to 1/wavelength[4]), and causes sky to assume its normally blue color.[6]

The most recent record of a blue sun and moon was in Sep-tember 1950, when a series of forest fires in British Columbia re-leased so much smoke into the sky that residents of eastern North America and much of western Europe saw a blue sun and moon.[7] The blue moon phenomenon is rare; you might say that it occurs only "once in a blue moon."[8]

The red tides and dead fish of 1884 are unrelated to the blue moon of 1883, except that in both events, occurring about a year apart, the seemingly unnatural colors are in fact perfectly nat-ural. The red tides of coastal Florida were caused by an enor-mous increase in the numbers of the dinoflagellate, *Gymnodinium brevis*, a microscopic photosynthetic organism. The ocean turned red during the day because the chlorophyll of this dinoflagellate emits red fluorescent light when nutrient limitations prevent photosynthesis from functioning normally.[9] These dinoflagel-lates produce toxins that are harmful to fish, birds, and people alike. The shellfish that feed on dinoflagellates are mostly unaf-fected, but they store dinoflagellate toxins in their bodies, so that anyone or anything eating these shellfish may suffer paralysis or even death.

A different dinoflagellate, *Gonyaulax polyedra*, is responsible for the red tides of southern California and actually emits blue bioluminescent light at night time (see sidebar, "Bioluminescence Basics"). Red tides from either *Gonyaulax* or *Gymnodinium* can cause significant ecological and economic damage. Their occur-rence is difficult to forecast and, once they occur, they are diffi-cult to control.[10] Unfortunately, red tides occur rather more often than once in a blue moon.

Bioluminescence Basics

Bioluminescence is the emission of light from biochemical reactions within a living organism. It occurs in many species, including bacteria, algae, plants, fungi, fireflies and other insects, fish, and jellyfish. In all these species, *luciferin* is the generic term for the substrate that produces the light, and *luciferase* is the term for the enzyme that catalyzes the reaction. Thus luciferin and luciferase do not refer to specific chemical or biological compounds. All known luciferases require molecular oxygen to oxidize a luciferin and form an enzyme-bound peroxide (any of several compounds with the -O-O- structure) as an intermediate.[*] The breakdown of this peroxide to a lower-energy compound results in the emission of energy in the form of light — bioluminescence. This reaction can be summarized as:

$$\text{Luciferin} + O_2 \rightarrow \text{Luciferase-Peroxide} \rightarrow \text{Luciferin}$$
$$\text{oxidation product} + \text{Light}$$

Among the many species that emit bioluminescence, there is considerable variation in the specifics of the biochemical reaction, the color of the light, the organs or cellular organelles from which the light originates, the biological and ecological function of the light, and the ancestry of the luciferase gene sequences. These disparate lines of evidence indicate that bioluminescence evolved independently many times.[†]

Although most people consider the dinoflagellates a nuisance, since 1958, *Gonyaulax polyedra* has been a popular model organism for research into biological rhythms. It turns out that the bioluminescence of *Gonyaulax polyedra* occurs at approximately twenty-four-hour intervals when it is held under constant environmental conditions in the laboratory.

What causes this rhythm of bioluminescence in *Gonyaulax*? What does *Gonyaulax* have to tell us about biological rhythms in general? A brief review of some of the biological rhythm research that preceded work with *Gonyaulax* provides the foundation for answering these questions.

Rhythm Research

At least since the time of the ancient Greeks, people have recognized that the leaves of many species of plants, especially those of the bean family (Fabaceae), assume a nearly horizontal position at day and a nearly vertical position at night.[11]

Jacques De Mairan, an eighteenth-century French astronomer, wondered whether these well-known leaf movements were controlled by changes in the external environment, such as sunlight, or by a mechanism within the plants themselves. In 1729 De Mairan performed a simple yet fundamental experiment on rhythmic leaf movements by using a species of mimosa, a member of the bean family.[12] When he placed mimosa plants in darkness, so that they never saw the light of day, he found that the plants maintained their daily changes in leaf position. Because bright sunlight was not required to maintain these daily leaf movements, De Mairan concluded that the rhythm was controlled by an endogenous mechanism.

De Mairan's line of research was pursued by many other biologists throughout the nineteenth century, including Charles Darwin and his son Francis, who devoted an entire book to plant movements, *The Power of Movement in Plants*, published in 1880 (see Chapter 9). In 1928 two young German botanists, Erwin Bünning and Kurt Stern, began their own studies of rhythmic leaf movements.[13] They discovered that dim light, which previous investigators had not controlled for and which we now know is detected by the plant pigment phytochrome (see Chapter 7 and figure A6, Appendix), controls the phase and period of the rhythmic leaf movements observed by De Mairan and subsequent researchers. When Bünning and Stern eliminated this dim light, they found that the "free-running" period of leaf movements (the period under constant conditions) was not exactly 24 hours, as measured by previous researchers, but rather 25.4 hours.[14] Because the earth's day-night cycle has a period of 24 hours, they considered their discovery as proof that rhythmic leaf movements are controlled by an endogenous biological clock. In other words, asynchrony of a biological rhythm and the earth's rhythm indicates that the biological rhythm is controlled by an endogenous factor. Such biological rhythms are now called "circadian" because they are about (*circa* in Latin) a day (*dies* in Latin) in length when measured under constant environmental conditions.

We now know that circadian rhythms occur not only in plants but also in fungi and animals — including humans. Circadian rhythms have long been known in single-celled organisms but only recently documented in bacteria, the simplest form of life.[15] Light-dark cycles can entrain, or alter the phase of, the circadian rhythms of all organisms, much like resetting an actual clock. But different species use different photoreceptive pigments for this entrainment. Animals apparently use rhodopsin, which consists of a vitamin A analog bound to a protein called opsin (see Chapter 2 and figure A1, Appendix), and a separate flavoprotein photoreceptor, which consists of a flavin (vitamin B-2) molecule bound to a protein.[16] Fungi also use a flavoprotein.[17] Plants use phytochrome, a tetrapyrole molecule bound to a protein, and a flavoprotein.[18] Significantly, entrainment of circadian rhythms by sunlight allows the circadian rhythms of all these disparate organisms to run with a period of exactly twenty-four hours in nature, a clear advantage for any organism that lives on this planet.

In summary, the general characteristics of circadian rhythms are:

- They have periods of about twenty-four hours under constant environmental conditions in the laboratory.
- They exhibit temperature compensation, in that they have approximately the same period at different temperatures.
- They can be phase-shifted by light and chemicals to have periods of exactly twenty-four hours in nature.[19]

Gonyaulax

As humans, we are certainly most interested in human circadian rhythms. Indeed, many scientists are actively studying human circadian rhythms.[20] Other scientists, however, study the circadian rhythms of various model organisms that are more amenable to experimental manipulation (see sidebar, "Model Organisms"). *Gonyaulax polyedra*, a dinoflagellate that emits bioluminescence at ~24-hour intervals, is a popular model organism for such studies because bioluminescence is an easily measured marker for the biological clock, and light has clear and easily measured effects on the rhythm of bioluminescence.[21]

Gonyaulax polyedra is a single-celled, saltwater species that is about 40 micrometers (0.04 millimeter) in diameter. Like other dinoflagellates (or "whirling flagellates"), *Gonyaulax* has two flagella, microscopic tail-like appendages, that beat in different planes and propel it through the ocean like a spinning top. Most dinoflagellates are covered by dense cellulose shells, with the shells of different species somewhat resembling the helmets of different medieval knights (see figure 10). Botanists once classified the dinoflagellates as algae because many species, including

Model Organisms

A rationale often given for studies of the circadian rhythms of model organisms (at least in many National Institutes of Health grant proposals) is that the mechanism is similar in all species, and that once we understand the mechanism in a model organism we will better understand the mechanism in humans. *Gonyaulax* is one well-known model organism for circadian rhythm research. Other commonly studied organisms include *Neurospora crassa,* a bread mold that produces conidiophores (spore-bearing structures) at ~24-hour intervals, and several species of *Drosophila,* fruit flies that undergo pupal eclosion (hatching), changes in locomotor activity, and other physiological processes at ~24-hour intervals.[‡]

All biological clocks apparently consist of negative-acting and positive-acting components (such as proteins) that interact in a feedback loop, with a delay imposed upon this interaction.[§] If circadian rhythms evolved independently several times, the clock components would be expected to differ among different species; analogously, the components of steam and diesel engines differ, even though steam and diesel trains have similar overt behavior.[#] Different organisms use different photoreceptive pigments to entrain their circadian rhythms, suggesting that circadian rhythms evolved independently in different organisms. Some genes that are known clock components, however (*wc-1* and *wc-2* in *Neurospora; per, cyc,* and *Clk* in *Drosophila; Clock, Per1, Per2, Per3,* and *bma1/mop3* in mice and presumably other mammals), share a common PAS domain in their sequence, suggesting a common derivation, possibly from an ancient photoreceptor gene.[°°] The identification and molecular analysis of circadian rhythms in additional species may answer the question of whether circadian clocks share common clock components or merely a common general mechanism.

Figure 10. Electron photomicrograph of *Gonyaulax polyedra*.
Gonyaulax polyedra is a bioluminescent, ocean-dwelling dinoflagellate that is about
0.04 millimeter in diameter and is covered by a dense cellulose shell.

Gonyaulax polyedra, are capable of photosynthesis. Recent molecular biology studies, however, suggest that dinoflagellates are
more closely related to apicomplexan parasites, a taxonomic
group that includes *Plasmodium,* the parasite that causes malaria.

It appears that dinoflagellates acquired their photosynthetic chloroplasts from an ancient algal endosymbiosis, a union in which a species of microscopic alga moved within a host cell and stayed there for millions of successive generations.[22] All chloroplasts have an endosymbiotic origin, though this has apparently happened many times throughout evolutionary history.

Gonyaulax normally emits bioluminescence at nighttime as many brief (0.1–0.5 second) flashes of blue light, with a steady glow occurring at the end of the night. Each cell emits about 240 million photons over the course of a day.[23] This light comes from scintillons, special organelles (membrane-enclosed structures within the cell) that are about 0.5 micrometer in diameter, about 1 percent of the diameter of the cell. The scintillons contain all the components necessary for bioluminescence: luciferin, the light-emitting compound (see figure A12, Appendix); luciferase, the enzyme that catalyzes oxidation of luciferin; and the luciferin-binding protein, a protein that binds to and stabilizes the inactive luciferin (see sidebar, "Bioluminescence Basics").[24]

At the beginning of each day, when the cells emit very little light, there is a tenfold reduction in the number of scintillons and their components (luciferin, luciferase, and luciferin-binding protein). At the beginning of each evening, a cell makes about four hundred new scintillons and synthesizes new luciferin, luciferase, and luciferin-binding proteins.[25] During the night, an electrochemical signal triggers an increase in H^+ concentration inside the scintillons. This causes activation of the luciferase enzyme and also causes the luciferin-binding protein to release the bound luciferin. Then luciferase oxidizes the luciferin and releases energy as bioluminescent light.

Clock Control

Bioluminescence is not the only rhythm in *Gonyaulax*. This tiny cell also has circadian rhythms in cell division, photosynthetic capacity, cell aggregation, and geotaxis (movement in response to gravity).[26] The molecular and biochemical changes that underlie these physiological rhythms, such as changes in the levels of proteins and metabolites, also exhibit circadian rhythms. Moreover, as with the bioluminescence rhythm, all these physiological rhythms can be reset by light.

It is conceivable that all the different *Gonyaulax* rhythms are controlled by the same clock, much in the way a clock-radio controls the hands of the clock and turns on the radio. Indeed, the rhythms for bioluminescence and cell aggregation normally have the same period but different phases — bioluminescence occurs at night and cell aggregation at day — suggesting control by the same clock. Under dim red light, however, the rhythms for bioluminescence and cell aggregation can be decoupled, with the rhythm for light flashes running progressively slower and the rhythm for aggregation running progressively faster. This indicates that *Gonyaulax* has at least two circadian clocks. Blue light is most important for regulating the bioluminescence rhythm, and red light is most important for regulating swimming behavior.[27]

Light absorption by photosynthetic pigments appears to be involved in resetting the bioluminescence circadian clock. Blue and red light are most effective in controlling photosynthesis and bioluminescence, and certain chemical inhibitors of photosynthesis also disrupt the bioluminescence rhythm.[28] When *Gonyaulax* is maintained under constant blue light, however, the bioluminescence period decreases as the level of light increases; under constant red light, the bioluminescence period increases as the level of light increases.[29] This indicates that two or more photoreceptors control the bioluminescence rhythm. Thus, just as *Gonyaulax* has two or more circadian clocks, it also has two or more photoreceptive pigments that control these clocks.

Adding to this complexity, researchers have recently shown that *Gonyaulax* cells contain melatonin, a pineal hormone that is best known for controlling the circadian rhythms of humans and other animals and for playing a role in seasonal affective disorder (see Chapter 5).[30] As in humans and other animals, melatonin levels in *Gonyaulax* increase at nighttime and decrease during the day. Moreover, the melatonin concentration in *Gonyaulax* is about the same as in the pineal gland of mammals, reaching a nighttime maximum of about 2.5 nanograms per milligram of protein. Although *Gonyaulax* researchers are still uncertain of the exact role of melatonin in the rhythms of this dinoflagellate, plant scientists have shown that melatonin occurs in plants from sixteen taxonomic families.[31] Thus it is possible that a great many organisms — animals, plants, algae, and dinoflagellates — use melatonin as a signal for the onset of darkness.

Once the molecular details of the circadian clocks of *Gonyaulax* are more fully understood, this knowledge may prove useful for treatment of the many human disorders that have underlying circadian or rhythmic components. These include jet lag, shift worker's syndrome, and seasonal affective disorder, as well as more serious medical conditions such as heart attacks and chronic myelogenous leukemia.[32] It is likely that there are differences in the molecular details of rhythm generation in *Gonyaulax* and in humans, but it seems likely that there are similarities in the underlying dynamical control mechanisms. An understanding of the circadian rhythms of *Gonyaulax* may even prove useful for predicting and controlling the noxious red tides of Florida and California.

Suggested Reading

Dunlap, J. C. (1999) Molecular bases for circadian clocks. *Cell* 96, 271–90.

Henry, M. (1998) Red Tide Links (www.mote.org/~mhenry/rtlinks.phtml).

Minnaert, M. J. G. (1974) *Light and Color in the Outdoors.* New York, Springer-Verlag (trans. and rev. L. Seymour, 1993).

Mittag, M., and J. W. Hastings (1996) Exploring the signaling pathway of circadian bioluminescence. *Physiologia Plantarum* 96, 727–32.

Roenneberg, T. (1996) The complex circadian system of *Gonyaulax polyedra*. *Physiologia Plantarum* 96, 733–37.

Scripps Institution of Oceanography (1998) Bioluminescence (siolib-155.ucsd.edu/mlatz/Biolum_intro.html).

Winfree, A. T. (1980) *The Geometry of Biological Time*. New York, Springer-Verlag.

14 Photosynthesis and the Great Salt Lake

We have some salt of our youth in us.

— Shakespeare

Observers of nature once classified all organisms as plants or animals. Beginning in the seventeenth century important discoveries by the Dutch biologist Antoni van Leeuwenhoek revealed that there was more to life. Between 1673 and 1723 Leeuwenhoek built microscopes from lenses that he painstakingly ground himself, and with these instruments he discovered animacules (little animals) — bacteria and other single-celled organisms. Leeuwenhoek submitted a series of letters to the Royal Society of London that described his important discoveries.[1] Many years later, biologists discovered that certain single-celled organisms lacked nuclei, the DNA-bearing structures found in the cells of all higher organisms. This led the French biologist Édouard Chatton to propose in 1937 that all organisms be classified as *eukaryotes*, whose cells have true (*eu* in Greek) nuclei (*karyotos*), or as *prokaryotes*, whose cells evolved before (*pro*) the nucleus.[2]

Modern biologists have an even greater appreciation of the diversity of life. We now know that organisms can be found in every imaginable environment on earth, from the boiling hot, sulfur-spewing vents that lie several miles beneath the ocean surface (see Chapter 2) to the dessicated interior of rocks in the frozen Antarctic.[3] Just as Leeuwenhoek showed that all organisms should not be classified as plants or animals, Carl Woese and colleagues showed that all prokaryotes should not be classified together. In 1977 Woese and colleagues reported their studies of the ribosomal RNA sequences of a variety of prokaryotes and eukaryotes and showed that the prokaryotes naturally separated into two distinct groups.[4] Based on these results, they proposed a tripartite classification of life into groups now called the Eukarya, single-celled and multicellular eukaryotic organisms; Eubacteria, prokaryotes which include the typical bacteria; and Archaea (or Archaebacteria, "primitive" bacteria), unusual methane-producing prokaryotes.[5]

Biologists now know much more about the Archaea. This large and diverse group includes the methanogens (methane-producing prokaryotes) studied by Woese, as well as extreme thermophiles, which inhabit environments where the temperature is typically about 90° C. (~200° F.), and halophiles, which inhabit highly saline environments.[6] Because most of the initially discovered Archaea lived in harsh environments, similar to those presumably present on the primordial earth, they were originally thought to have evolved before the Eubacteria or Eukarya. Woese and many other biologists, however, while accepting the tripartite classification of life, now believe that the Eubacteria are the most primitive group.[7] According to this view, an ancestor of the Archaea and Eukarya evolved from primitive Eubacteria and then the Archaea and Eukarya diverged from one another (see figure 11).[8] Thus humans (which are Eukarya) are more closely related to the Archaea than to Eubacteria.

Halobacterium

Clearly, discovery of the Archaea has provided a lot of excitement to evolutionary biologists. One species of Archaea, *Halobacterium salinarum,* has also excited many photobiologists and computer scientists (see sidebar, "Computing with Bacteriorhodopsin").[9] This species uses a unique form of photosynthesis to

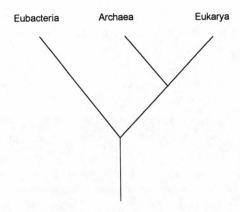

Figure 11. Evolutionary relationships of
Archaea, Eubacteria, and Eukarya. Many
biologists now support the classification of all
organisms into a tripartite system, in which
the Archaea are more closely related to
Eukarya than Eubacteria.

transform the energy of sunlight into chemical energy that it uses
for growth and reproduction. It also has two unique photosensitive pigments that provide the cells with information about their
light environment. In short, *Halobacterium* uses sunlight as a
source of energy and as a source of information.

Halobacterium salinarum is a rod-shaped cell that is typically
two to three micrometers long and half a micrometer wide. One
or two bundles of flagella (whiplike appendages) that derive
from the cell poles move the cells around, with molecular motors
at the base of the flagella controlling their direction of rotation
and the direction of cell movement.[10] *Halobacterium* thrives in
brackish waters, such as the Great Salt Lake of Utah and the
Dead Sea of the Middle East, that have salt concentrations of
about 4 to 5 Molar (about 25 percent salt by weight) and smell a

Computing with Bacteriorhodopsin

One of the main reasons for the constant increase in computer speed over the past fifty years has been improvement in lithography, the technique used to etch circuits onto computer chips.[*] As lithography improves, the size of the logic gate, the computer's basic circuit element that switches between two states (0 and 1), gets smaller. A chip with smaller gates means that signals can travel faster between gates and computers can run faster. Just how small can a gate get?

Recent research has shown that it is possible to use a single molecule of bacteriorhodopsin, the photosensitive protein of *Halobacterium,* as a computer switch.[†] Bacteriorhodopsin is much smaller than the logic gates used in modern computers. It is easy to envision bacteriorhodopsin as a computer switch, because flashes of light from a laser beam can change the protein between two states in an easily controlled and predictable manner, and these two states can be measured with another laser beam.

An additional advantage of a bacteriorhodopsin-based computer is that chips can be connected in three-dimensional arrays.[‡] Because a two-dimensional laser array can read and write planes of data within a three-dimensional structure, such a computer will have much faster data transfer rates. Silicon-based chips must be placed on two-dimensional "wafers" because transistors in the middle of a three-dimensional structure would rapidly overheat and break down. A three-dimensional bacteriorhodopsin-based storage device with two-dimensional arrays of lasers is now under development by Robert Birge and colleagues at Syracuse University.[§] Although the development of a purely biomolecular computer is probably far off, Birge predicts that a silicon-protein hybrid computer will become available in the near future.[#]

bit like dirty old sneakers. By comparison, the salinity of the ocean is only about 0.6 Molar (about 3 percent salt by weight). While most other cells will die in the water where *Halobacterium* thrives, *Halobacterium* cells will die if grown in solutions that contain less than about 20 percent salt by weight.[11] When a large mass of *Halobacterium* cells accumulate, they typically turn their resident lake or salt lick to a red-purple color. Travelers who fly into San Francisco often notice that the salt flats below, which are populated with *Halobacterium*, are purple.[12]

Ten years before Carl Woese and colleagues proposed their tripartite classification system and placed *Halobacterium* within the Archaea, Walther Stoeckenius and colleagues showed that the membranes of *Halobacterium* could be broken apart and separated into three fractions: a yellow fraction, a red fraction, and a purple fraction.[13] The yellow fraction contains the protein-rich walls of gas vacuoles that allow the cells to adjust their depth in the water. The red fraction, which contains orange carotenoid pigments (see Chapter 15) and red porphyrin pigments (see Chapter 6), also has the enzymes necessary for cellular respiration and the synthesis of ATP (adenosine triphosphate), an important high-energy compound for cells. The quantity of the purple fraction varies dramatically according to growth conditions. Cells grown in darkness or dim light with abundant oxygen have very little of the purple pigment. In contrast, cells grown in bright light and under low concentrations of oxygen accumulate oval, purple-colored patches that are about one micrometer in diameter and occupy more than half the surface of the membrane.

Initial studies of these purple patches showed that they were composed primarily of a protein in a crystalline array, with about 20,000 molecules per array. Later studies showed that this protein traverses the membrane and that, like the rhodopsin in our own eyes (see Chapter 1), each protein molecule is attached to a molecule of retinal, a vitamin A analog, by a special bond called a "protonated Schiff base." The retinal gives bacteriorhodopsin its characteristic purple color, since it strongly absorbs orange-red light (maximal absorption at 568 nm).[14] This bacterial rhodopsin-like protein was appropriately named "bacteriorhodopsin."

Photosynthesis Without Chlorophyll

Initial physiological studies showed that when *Halobacterium* cells were starved and maintained in an oxygen-free environment, they still made ATP in the presence of bright light. Because cellular respiration cannot occur under these conditions, *Halobacterium* cells must be able to harness the energy in sunlight to synthesize ATP. The wavelengths of light most effective in driving ATP synthesis are the same wavelengths that are most strongly absorbed by bacteriorhodopsin, implicating bacteriorhodopsin as the photoreceptive pigment.[15] In other words, *Halobacterium* cells, which completely lack chlorophyll, use bacteriorhodopsin to perform a unique form of photosynthesis.

In 1971 Stoeckenius and Dieter Osterheldt showed that bacteriorhodopsin undergoes a cycle when it absorbs light.[16] In particular, light rapidly transforms the "linear" all-trans retinal of bacteriorhodopsin into the "bent" 13-cis form of retinal, which then slowly returns to all-trans retinal (see figure A13, Appendix). During each cycle, bacteriorhodopsin pumps a proton out of the cell; later in the cycle, it takes up a proton from inside the cell.[17] This movement of protons out of the cell creates a pH gradient, with the inside of the cell becoming more alkaline and the outside more acidic. The resulting pH gradient is used to generate ATP from adenosine diphosphate (ADP) as protons flow back into the cell through ATPase (see figure 12). This process, termed chemiosmosis, occurs in all cells and can be considered a "biological proton battery," with the inside and outside of the cell considered as the two compartments, the ATPase as a combina-

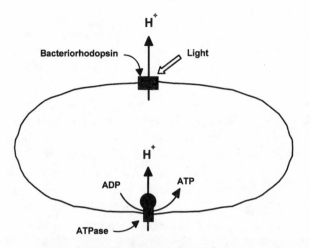

Figure 12. Photosynthesis in Halobacteria.
Light absorbed by bacteriorhodopsin pumps protons from
inside to outside the cell. These protons are then used by
the respiratory complex (ATPase) to synthesize adenosine
triphosphate (ATP), a high-energy compound used to drive
many chemical reactions in the cell.

tion of platinum wire and hydrogen gas, the flow of protons as a
flow of current, and the work performed as the synthesis of ATP
from ADP.[18]

Modern chemists use bacteriorhodopsin as a model for
study of the chemical dynamics of proteins.[19] This important
field studies the enzymes that drive all the chemical reactions in
living organisms by examination of the formation and breaking
of the chemical bonds of proteins. Because protein chemical
dynamics occur on a picosecond (10^{-12} second) time scale,
chemists must use very special techniques for these studies. All
such methods to date employ the "pump-probe" technique, in

which a very brief pulse of light (the "pump") triggers chemical changes in a protein that are studied by various measurement methods (the "probe").[20] Because this method must employ proteins that are activated by light, chemical dynamics studies can be performed only on a very small number of all the proteins found in nature. Bacteriorhodopsin is an important model protein for understanding the general features of the chemical dynamics of enzymes because many other proteins found in microbes, animals, plants, and fungi also function as ion pumps.[21]

In the late 1970s, experiments performed with a special mutant of *Halobacterium* that completely lacked bacteriorhodopsin showed that light still promoted ATP synthesis.[22] This surprising result was attributed to a new light-sensitive protein called halorhodopsin. Halorhodopsin was originally thought to be a light-driven sodium pump, but in 1982 it was shown to be a light-activated protein that pumps chloride ions into the cell.[23] Halorhodopsin is similar to bacteriorhodopsin in that it is a retinal-based protein, but bacteriorhodopsin uses the energy of light to expel protons from the cell, whereas halorhodopsin uses the energy of light to take up chloride ions.

Why does *Halobacterium* have this additional photosensitive protein and why does it pump chloride ions into the cell? In its typical habitat, the sodium chloride (NaCl) concentration outside the cell is very high; this is balanced by a very high internal concentration of potassium chloride (KCl). The cells normally expel sodium ions and take up potassium ions to maintain this balance. *Halobacterium* cells must take up potassium ions more rapidly than they expel sodium ions in order to grow (take up water and increase in volume). As the cells take up potassium ions, they must also take up chloride ions; this is an energy-requiring process because the inside of the cell is more negatively charged than the outside.[24] Thus halorhodopsin uses the energy of sunlight to drive the important process of chloride ion uptake so that cells can grow.

Sensory Rhodopsins

In 1982 researchers at the University of Texas discovered that light controlled the movement of special *Halobacterium* mutants that lacked bacteriorhodopsin and halorhodopsin.[25] This indicated that cells must have one or more additional photoreceptive pigments. Subsequent studies showed that the cells used two retinal-based pigments, subsequently named sensory rhodopsin-I and sensory rhodopsin-II, to control swimming away from ultraviolet and blue light and toward orange-red light (see figure 13).[26] This behavior makes sense because bacteriorhodopsin and halorhodopsin use orange-red light for photosynthesis, and DNA and other molecules can be damaged by ultraviolet radiation (see Chapter 4). Unlike bacteriorhodopsin and halorhodopsin, but like the rhodopsin of our own eyes (see Chapter 1), these new pigments are sensory rhodopsins. In other words, they provide cells with information about the light environment, not energy in the form of ATP.

How do *Halobacterium* cells use their four rhodopsins to flourish in their natural habitat? Unfortunately, while *Halobacterium* is extensively studied in the laboratory, little information is available on the life of these cells in nature. The extensive physiological studies suggest an answer, however.[27] When oxygen and carbohydrates or other food sources are readily available, the cells grow normally and move away from sunlight to prevent damage caused by photooxidation. Sensory rhodopsin-II, which strongly absorbs blue-green light, controls this response and is the only rhodopsin present in significant amount when cells are grown under these conditions. In the dark, the cells move about randomly, performing what can be considered a "three-dimensional random walk."[28] When the oxygen level decreases, the cells no longer make sensory rhodopsin-II; they begin to make sensory rhodopsin-I, bacteriorhodopsin, and halorhodopsin, and they begin to photosynthesize. Although the cells cannot survive

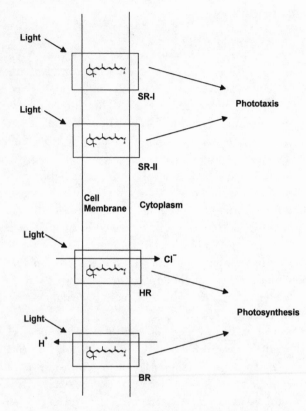

Figure 13. The four rhodopsins of *Halobacterium*.
All four *Halobacterium* rhodopsins are membrane-bound proteins that contain all-trans retinal. Sensory rhodopsin I (SR-I) and sensory rhodopsin II (SR-II) control phototaxis, movement toward and away from light. Bacteriorhodopsin (BR) and halorhodopsin (HR) drive photosynthesis by pumping protons out of the cell and chloride ions into the cell, respectively.

by sunlight alone, light helps them survive through periods when they are exposed to very low levels of oxygen. Sensory rhodopsin-I controls movement of the cells toward red-orange light, the same color that is most effectively absorbed by bacteriorhodopsin and halorhodopsin. In addition, a special form of sensory rhodopsin-I (a long-lived intermediate called S_{373}) controls movement away from ultraviolet radiation. Thus the cells move toward orange-red light to maximize photosynthesis but away from ultraviolet radiation to avoid cellular damage. Further complicating the apparently distinct roles of the four photoreceptive pigments, recent studies show that bacteriorhodopsin also plays a role in positive phototaxis under high levels of light.[29]

Although the bacteriorhodopsin of *Halobacterium* and the rhodopsin in the eyes of humans and other animals both consist of retinal (a vitamin A analog) bound to a protein, these proteins are apparently unrelated.[30] There is another fascinating distinction between bacteriorhodopsin and the rhodopsin of our own eyes.[31] Our rhodopsin uses the unstable "bent" (11-cis) retinal to absorb light; light triggers vision by changing this into the more stable "straight" (all-trans) retinal and energy is used to regenerate the "bent" form (see figure A1, Appendix).[32] In contrast, the bacteriorhodopsin of *Halobacterium* uses "straight" retinal to absorb light; light drives photosynthesis by changing this into "bent" (13-cis) retinal and this reverts spontaneously to the "straight" form (see figure A13, Appendix).

It seems remarkable that *Halobacterium* uses a retinal bound to a protein to drive photosynthesis and that humans use a retinal bound to an unrelated protein to trigger vision. Max Delbrück first remarked in 1976 that there are few types of photosensory and energy transducing pigments found in nature, and he called this the "paradox of the short list."[33] It is still not clear why nature uses so few types of photosensory and energy transducing pigments, but it calls to mind a famous remark by Einstein: "Things should be made as simple as possible, but not simpler."[34]

Suggested Reading

Hoff, W. D., et al. (1997) Molecular mechanism of photosignaling by archael sensory rhodopsins. *Annual Review of Biophysics and Biomolecular Structure* 26, 223–58.

Lanyi, J. K. (1997) Mechanism of ion transport across membranes. Bacteriorhodopsin as a prototype for proton pumps. *Journal of Biological Chemistry* 272, 31209–17.

Membrane Proteins of Known Three-Dimensional Structure (blanco. biomol.uci.edu/Membrane_Proteins_xtal.html).

15 Too Much of a Good Thing

Can one desire too much of a good thing?
— *Shakespeare*

Just as the wolves eat the moose and the moose eat the plants, each class of organisms in a food chain consumes the food energy of a different class of organisms. All food chains require primary producers — organisms that provide an energy input. Although there are many different types of food chains on earth, there are only a few basic mechanisms for putting energy into food chains. One of the most unusual of these is used by bacteria that live near hydrothermal vents that lie a mile or more beneath the ocean surface. They use energy from the oxidation of hydrogen sulfide, which is toxic to most organisms, to convert carbon dioxide into glycogen (see Chapter 2).

Photosynthesis, the biological conversion of light energy into chemical energy, is the most important mechanism for putting energy into food chains, and plants and algae are the most important primary producers on earth. According to one widely quoted estimate, the photosynthesis in all the terrestrial and aquatic ecosystems on earth produces about 187 billion tons of carbohydrates per year.[1] This corresponds to roughly 31 tons of carbohydrate per year for each person now living on earth.[2]

This is an enormous amount of food energy, though only a small fraction of the total amount of light energy that strikes the earth each year. On a clear day at noontime, roughly 10^{21} photons of photosynthetically active radiation strike each square meter of the earth every second (see table 1, p. 6), but plants use far less than 1 percent of the energy in these photons to make carbo-hydrates.[3] Laboratory experiments have shown that the photo-synthesis of exposed leaves on certain sun-loving species, such as corn and sugar cane, continues to increase as the light level increases even beyond the highest levels found in nature. The photosynthesis of exposed leaves on other sun-loving species, such as cotton and wheat, saturates near the level present on a clear day. In contrast, the photosynthesis of exposed leaves on most shade-loving species, such as the spider plant, philoden-dron, and other familiar house plants, is saturated by light levels less than 10 percent of that present on a clear day.[4]

What happens when plants get too much sunlight — that is, when they get too much of a good thing?

Too Much Light

One of the worst things that happens when plants get too much light is "photooxidative stress," the light-driven generation of toxic forms of oxygen.[5] The most biologically significant toxic forms of oxygen are singlet oxygen ($^1O_2$), the hydroxyl radical ($OH\cdot$), hydrogen peroxide (H_2O_2), peroxynitrate ($ONOO^-$), and hypohalite (OCl^-).[6] These compounds are easily intercon-verted, so it is often difficult for biologists to ascribe a certain type of cell damage to one particular form. Nonetheless, most forms are harmful because they can destroy essential biological molecules such as lipids (the structural components of cell mem-branes), amino acids (the structural components of proteins), and nucleotides (the structural components of RNA and DNA).[7] Toxic forms of oxygen can even destroy chlorophyll, the very pigment responsible for their formation.

There is growing evidence that toxic forms of oxygen, al-beit generated by different mechanisms, also have detrimental effects on human health. This has led physicians and nutritionists to recommend inclusion of antioxidants (vitamin C, vitamin E, flavonoids, and carotenoids) in the diet as protection against cancer, heart disease, and other ailments.[8]

In plants that receive too much light, toxic forms of oxygen are generated mostly by two separate series of reactions: one generates singlet oxygen, and the other generates superoxide and hydrogen peroxide (see sidebar, "Generation of Toxic Forms of Oxygen in Plants").

Singlet oxygen generation. Normally, when chlorophyll absorbs sunlight, it is excited to a short-lived high-energy level, called the singlet state. The energy in singlet-state chlorophyll is passed to special reaction centers, where it drives photosynthesis and the conversion of carbon dioxide to carbohydrate.[9] Under very bright light, singlet chlorophyll can accumulate and relax to a longer-lived high-energy level called the triplet state.[10] Triplet-state chlorophyll can pass its energy to oxygen, forming singlet oxygen, a toxic form of oxygen.[11] Coincidentally, a similar mechanism causes skin photosensitivity in people with certain forms of porphyria, a group of heritable metabolic diseases (see Chapter 6).

Superoxide and hydrogen peroxide generation. Normally, after chlorophyll has absorbed light and passed its energy to special reaction centers, this energy is used to split apart water molecules. As the oxygen from the water is released, electrons pass through a series of special electron carriers and ultimately form NADPH (reduced nicotinamide adenine dinucleotide phosphate), an important high-energy compound that plants use to convert carbon dioxide into carbohydrate.[12] When the light is very bright, however, ferredoxin, one of the electron carriers, can pass electrons to oxygen and generate superoxide.[13] Superoxide is not particularly toxic, but it is easily converted to other toxic forms of oxygen, such as hydrogen peroxide.

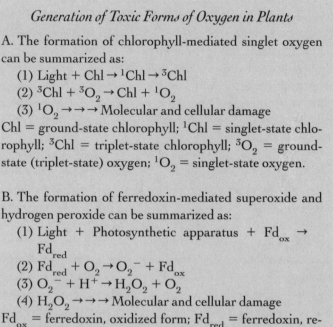

Generation of Toxic Forms of Oxygen in Plants

A. The formation of chlorophyll-mediated singlet oxygen can be summarized as:

(1) $\text{Light} + \text{Chl} \rightarrow {}^1\text{Chl} \rightarrow {}^3\text{Chl}$

(2) ${}^3\text{Chl} + {}^3\text{O}_2 \rightarrow \text{Chl} + {}^1\text{O}_2$

(3) ${}^1\text{O}_2 \rightarrow \rightarrow \rightarrow \text{Molecular and cellular damage}$

Chl = ground-state chlorophyll; ${}^1\text{Chl}$ = singlet-state chlorophyll; ${}^3\text{Chl}$ = triplet-state chlorophyll; ${}^3\text{O}_2$ = ground-state (triplet-state) oxygen; ${}^1\text{O}_2$ = singlet-state oxygen.

B. The formation of ferredoxin-mediated superoxide and hydrogen peroxide can be summarized as:

(1) $\text{Light} + \text{Photosynthetic apparatus} + \text{Fd}_{ox} \rightarrow \text{Fd}_{red}$

(2) $\text{Fd}_{red} + \text{O}_2 \rightarrow \text{O}_2^- + \text{Fd}_{ox}$

(3) $\text{O}_2^- + \text{H}^+ \rightarrow \text{H}_2\text{O}_2 + \text{O}_2$

(4) $\text{H}_2\text{O}_2 \rightarrow \rightarrow \rightarrow \text{Molecular and cellular damage}$

Fd_{ox} = ferredoxin, oxidized form; Fd_{red} = ferredoxin, reduced form; O_2 = oxygen; O_2^- = superoxide; H^+ = proton; H_2O_2 = hydrogen peroxide.°

Coping with Too Much Light

Plants have a variety of mechanisms for coping with too much light. Many plants acclimate to high-light levels by making anatomical changes that reduce the possibility of damage.[14] The leaves of plants grown in full sun, for example, are often smaller and thicker, coated with thin layers of wax and pubescence, and oriented with the long axis parallel to the sun's rays. All these changes decrease light absorption by chlorophyll and reduce the damage from excessive light.

But these structural modifications are not entirely effective. An additional mechanism is the removal of the toxic forms of oxygen once they have formed by the Mehler-peroxidase reaction.[15] This special set of reactions, driven by the enzymes superoxide dismutase and ascorbate peroxidase, remove superoxide and hydrogen peroxide and produce water as a harmless byproduct. Plants do not have enzymes that can remove singlet oxygen ($^1O_2$) or the hydroxyl radical ($OH\cdot$).[16] Instead, they synthesize ascorbate (vitamin C), glutathione, flavonoids, alpha-tocopherol (vitamin E), carotenoids, and other compounds that can quench the high energy present in these toxic compounds. Thus carotenoids (see sidebar, "Carotenoids") are important plant pigments not simply because they give red, orange, yellow, or purple pigmentation to different parts of a plant but also because they can inactivate toxic forms of oxygen.[17]

An additional defense strategy that plants use is removal of the excitation energy from chlorophyll before it can be used to generate toxic forms of oxygen. This is exactly what the xanthophyll cycle does (See figure A14, Appendix). Xanthophylls and carotenes are the two classes of carotenoids found in plants. The xanthophyll cycle can be compared to a lightning rod, in that it harmlessly removes excess light energy that is absorbed by chlorophyll.[18] Under high levels of light the xanthophyll cycle de-excites chlorophyll and harmlessly dissipates this energy as heat. Apparently, the excess light energy that is absorbed by chlorophyll is transferred directly from light-activated (singlet) chlorophyll to zeaxanthin and then harmlessly dissipates as heat.[19]

How does the xanthophyll cycle protect plants from excess light? Under high levels of light, violaxanthin changes into antheraxanthin and then to zeaxanthin; zeaxanthin removes the energy in excited chlorophyll.[20] Under low levels of light, zeaxanthin is transformed to antheraxanthin and then to violaxanthin. Because of its different chemical structure, violaxanthin is unable to accept excitation energy from chlorophyll. The overall cycle is

Carotenoids

The carotenoids constitute a large group of naturally occurring, fat-soluble pigments that are red, orange, yellow, or purple, depending on the molecular structure.[†] There are hundreds of different carotenoids in nature, and they are well known for giving color to the fruits, flowers, and roots of higher plants. Carotenoids and flavonoids (another class of pigments) are mainly responsible for the color of autumn leaves.[‡]

The two main classes of carotenoids are the carotenes, pure hydrocarbon molecules that consist only of hydrogen and carbon, and the xanthophylls, oxygenated derivatives of the carotenes that typically have two or more oxygen atoms per molecule. Typically carotenoids are synthesized from eight 5-carbon isoprene molecules and contain forty carbon atoms.[§]

Higher plants, algae, fungi, and bacteria can synthesize carotenoids. In addition, many animals that eat carotenoid-containing plants or algae accumulate carotenoids in certain tissues. For example, salmon flesh, lobster shells, and flamingo feathers are red because the algae in the diets of these organisms contain carotenoids. Significantly, humans and other animals use carotenes to synthesize vitamin A, an essential vitamin and the precursor of retinal, the light-sensitive component of rhodopsin, the pigment used in animal vision (see figure A1, Appendix). Zeaxanthin has an important photoprotective role in the human eye in that it prevents age-related macular degeneration, a disease that can reduce visual acuity or even cause blindness.[#]

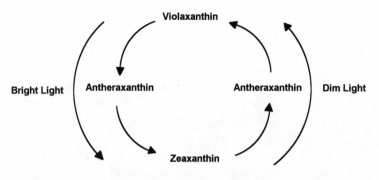

Figure 14. The xanthophyll cycle.
Plants accumulate zeaxanthin in bright sunlight and violaxanthin in deep
shade. Zeaxanthin protects plants by removing the energy in light-excited
chlorophyll so that it does not cause photooxidative stress. Violaxanthin is
unable to accept excitation energy from light-excited chlorophyll.

summarized in figure 14. Under stressful conditions, certain
evergreens with thick hard leaves (known to ecologists as sclero-
phytes) maintain high levels of zeaxanthin continuously, night
and day. This appears to be important for the acclimation of these
plants to winter conditions.[21]

Because zeaxanthin protects plants from the damage
caused by excess light, why don't plants simply maintain high
levels of zeaxanthin at all times? If this occurred, zeaxanthin
would always be "stealing" some of the energy from light-activated chlorophyll, and this energy could not be used to drive photosynthesis. In fact, within minutes of transferring a plant from
bright light to dim light, the leaves transform zeaxanthin into violaxanthin so that zeaxanthin does not compete with photosynthesis for the excitation energy of chlorophyll; within minutes of
transferring a plant from dim light to bright light, the leaves
rapidly transform violaxanthin into zeaxanthin as protection
against photooxidative damage.[22] Thus at dawn and dusk violax-

anthin levels are high, and antheraxanthin and zeaxanthin levels are low; at midday, violaxanthin levels are low and zeaxanthin levels are high.[23] In fact, within a single plant, the east-facing leaves have higher zeaxanthin levels in the morning and the west-facing leaves have higher zeaxanthin levels in the afternoon.[24] The leaves of plants grown in full sun also have a greater total concentration of xanthophyll cycle compounds (violaxanthin + antheraxanthin + zeaxanthin).[25]

Just how important are carotenoids and the xanthophyll cycle in protecting plants from excess light? Carotenoids appear to be very important, because application of norflurazon, a herbicide that blocks synthesis of carotenoids, effectively kills all plants exposed to bright light.[26] The xanthophyll cycle also appears to be very important, because it is present in all major families of land plants, suggesting that it arose very early in the evolution of plants.[27] Presumably, nearly all plants experience excess light at some point in their lives, and they all need a way to dissipate this energy to prevent formation of toxic oxygen derivatives.

Perhaps the best way to determine the true importance of the xanthophyll cycle is to study plants in which this cycle is specifically disrupted. This is analogous to determining the function of a car's radiator by draining it of water and observing the effect upon the car's overall performance. Plant biologists have isolated specific mutants of *Arabidopsis thaliana* (mouse ear cress), a little plant in the mustard family and a much studied model organism, that are unable to accumulate zeaxanthin in excess light (*npq1* mutants) and other mutants that accumulate zeaxanthin under all conditions (*npq2* mutants).[28] Studies of the *npq1* mutants showed that zeaxanthin is indeed responsible for most of the dissipation of excess light energy from chlorophyll; the *npq2* mutants, however, which had abnormally high levels of zeaxanthin, were no more protected from excess light than normal plants.

To complement these studies with *Arabidopsis,* a land plant, the same researchers performed studies with mutants of *Chlamydomonas reinhardtii,* a single-celled organism that is one of the most common freshwater, photosynthetic, green algae.[29] In contrast to the results with *Arabidopsis, Chlamydomonas* mutants that are unable to synthesize zeaxanthin show the same growth rate as normal cells, indicating that the xanthophyll cycle plays a less essential role in this species. Another *Chlamydomonas* mutant *(lor1)* that is unable to synthesize lutein, a different carotenoid, also shows normal growth.[30] *Chlamydomonas* double mutants *(npq1 lor1)* that can synthesize neither zeaxanthin nor lutein, however, have drastically reduced growth rates.

These important mutant studies demonstrate that the xanthophyll cycle has different roles in protecting land plants and algae from high levels of light. In *Arabidopsis* (and presumably other land plants), the xanthophyll cycle is essential for protection against excess light; in *Chlamydomonas* (and presumably other green algae) the xanthophyll cycle is used in conjunction with other carotenoids for protection against excess light. Moreover, *Chlamydomonas* swims downward when the sun is too bright.

While physiologists, ecologists, biochemists, and molecular biologists continue to study how plants protect themselves from excess light, and genetic engineers attempt to design plants that are better able to withstand the stress of excess light, I find it interesting to contemplate what a truly tangled bank the earth would be if plants used all the sunlight they absorb to drive photosynthesis, or even if they were not harmed by excess light.[31] Photosynthesis is now responsible for making about 31 tons of carbohydrate per year for each person living on the earth. What if they made twice as much? How would this change the war of nature, the battle of famine and death? Would humans and other higher animals have evolved under such a scenario? There is grandeur in this view of life, but there is also grandeur in life as it is, with its saturation of photosynthesis, its toxic derivatives of oxygen, its antioxidants, its xanthophyll cycle, and its plant biologists who study such things.

Suggested Reading

Cadenas, E. (1989) Biochemistry of oxygen toxicity. *Annual Review of Biochemistry* 58, 79–110.

Demmig-Adams, B., and W. W. Adams (1996) The role of xanthophyll cycle carotenoids in the protection of photosynthesis. *Trends in Plant Science* 1, 21–26.

Niyogi, K. K., et al. (1997) *Chlamydomonas* xanthophyll cycle mutants identified by video imaging of chlorophyll fluorescence quenching. *Plant Cell* 9, 1369–80.

———— (1998) *Arabidopsis* mutants define a central role for the xanthophyll cycle in the regulation of photosynthetic energy conversion. *Plant Cell* 10, 1121–34.

Appendix
A Menagerie of Molecules

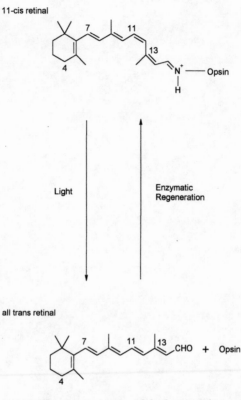

Figure A1. Photoisomerization of rhodopsin (Chapter 1).
Photoisomerization (light-induced structural change) of
rhodopsin is the primary event of vision in the many
organisms that have rhodopsin-based vision (P. Buser and
M. Imbert, 1995, *Vision*, Cambridge, MIT Press). The
rhodopsin of humans uses the "bent" 11-cis-retinal as the
light-absorbing component, which is bound to the protein
opsin. Upon light absorption, rhodopsin changes to the
"straight" all-trans-retinal and free opsin. A series of
enzymatic reactions regenerates rhodopsin. Compare this
reaction with the photocycle of bacteriorhodopsin (see
figure A13 below).

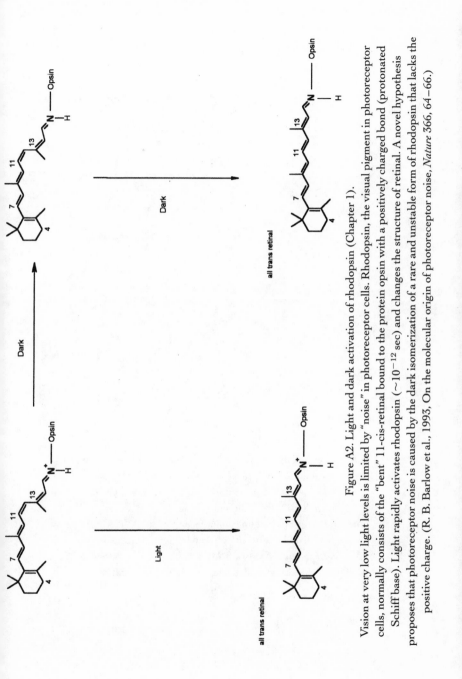

Figure A2. Light and dark activation of rhodopsin (Chapter 1).

Vision at very low light levels is limited by "noise" in photoreceptor cells. Rhodopsin, the visual pigment in photoreceptor cells, normally consists of the "bent" 11-cis-retinal bound to the protein opsin with a positively charged bond (protonated Schiff base). Light rapidly activates rhodopsin ($\sim 10^{-12}$ sec) and changes the structure of retinal. A novel hypothesis proposes that photoreceptor noise is caused by the dark isomerization of a rare and unstable form of rhodopsin that lacks the positive charge. (R. B. Barlow et al., 1993, On the molecular origin of photoreceptor noise. *Nature* 366, 64–66.)

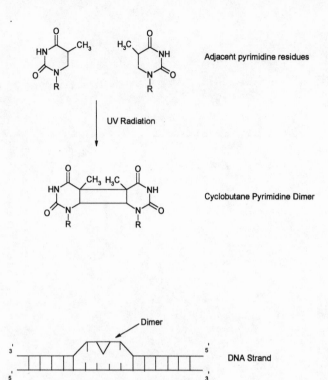

Figure A3. Cyclobutane pyrimidine dimer (Chapter 4).
Ultraviolet radiation is harmful to most organisms because it can
damage their DNA. The most common type of DNA damage
caused by ultraviolet radiation is the cyclobutane pyrimidine
dimer (CPD) (S. E. Freeman et al., 1989, Wavelength
dependence of pyrimidine dimer formation in DNA of human
skin irradiated in situ with ultraviolet light, *Proceedings of the
National Academy of Sciences, USA* 86, 5605–9). The formation of
the CPD is shown in detail at top and diagrammatically at
bottom.

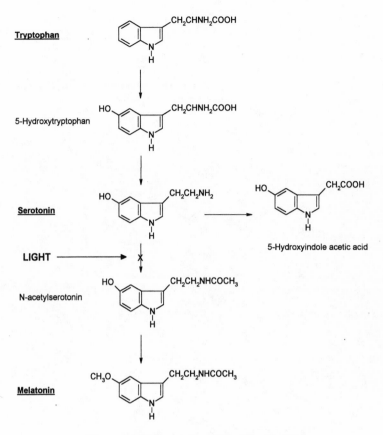

Figure A4. Melatonin synthesis (Chapter 5).

Disruption in the levels of melatonin and serotonin, compounds synthesized in the pineal gland, has been implicated in causing SAD, seasonal affective disorder (M. Shafii and S. L. Shafii, 1990, *Biological Rhythms, Mood Disorders, Light Therapy, and the Pineal Gland*, Washington, D.C., American Psychiatric Press). The amino acid tryptophan (top) moves from the blood stream into the pineal gland and is converted to serotonin and then melatonin (bottom). During the day, light blocks norepinepherine release, inhibiting melatonin synthesis and increasing the level of serotonin. At night, nerve fibers connected to the pineal gland release norepinepherine, stimulating the synthesis of melatonin and decreasing the level of serotonin. Melatonin is transported into the bloodstream and is eventually broken down in the liver and then excreted.

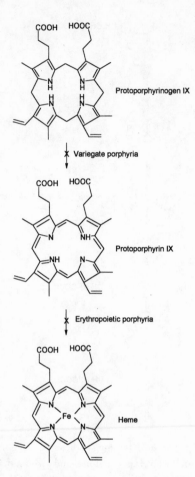

Figure A5. Final steps of heme synthesis (Chapter 6).

Heme is a red, iron-containing pigment that is an important component of hemoglobin, the oxygen-carrying protein of red blood cells. The last two steps of heme synthesis are conversion of protoporphyrinogen IX to protoporphyrin IX, and of protoporphyrin IX to heme. Several researchers have suggested that George III (1738–1820), King of England, Scotland, and Ireland, suffered from variegate porphyria, a disease in which the gene for synthesis of protoporphyrin IX (protoporphrinogen oxidase) is defective (J. C. G. Röhl et al., 1998, *Purple Secret*, London, Bantam). See table A1 below.

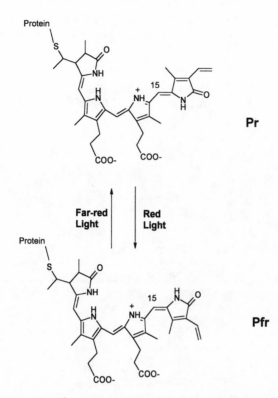

Figure A6. Photoisomerization of phytochrome (Chapter 7).
A red/far-red light reversible photoreaction controls many physiological
and developmental responses in plants. The photoisomerization (light-
induced structural change) of the open-chain tetrapyrole of phytochrome
at the 15,16 double bond is responsible for this effect (D. L. Farrens et al.,
1989, Surface-enhanced resonance Raman scattering spectroscopy applied
to phytochrome and its model compounds, part 2, Phytochrome and
phycocyanin compounds, *Journal of the American Chemical Society* 111, 9162–
69). Red light, which is strongly absorbed by the form of phytochrome
designated Pr, causes accumulation of the far-red absorbing form of
phytochrome (Pfr); far-red light, which is strongly absorbed by Pfr,
causes accumulation of Pr. In the semiextended conformations of the
two forms of phytochrome depicted here, note that the rightmost
pyrole ring "flips" back and forth.

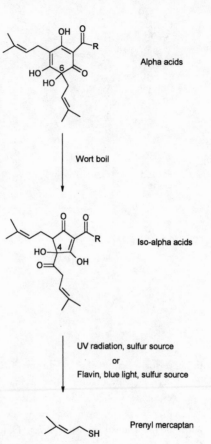

Figure A7. Sunstruck reaction (Chapter 8).

In making beer, a brewer boils the sweet wort with hops prior to
fermentation. This transforms the alpha-acids of hops into bitter-tasting
iso-alpha-acids, which balance the residual sweetness in the fermented
beer. Exposure of iso-alpha-acids to ultraviolet radiation leads to the
formation of prenyl mercaptan, a "skunky" smelling compound.
(J. Templar et al., 1995, Formation, measurement and significance of
lightstruck flavor in beer: A review, *Brewers Digest,* May, pp. 18–25.)

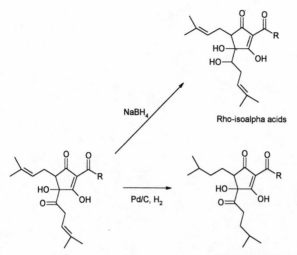

Rho-isoalpha acids

Isoalpha acids

NaBH$_4$

Pd/C, H$_2$

Tetra-hydroisoalpha acids

NaBH$_4$

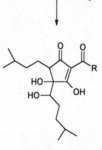

Hexa-hydroisoalpha acids

Figure A8. Photostable hop compounds (Chapter 8).
Chemists can prepare numerous bitter-tasting compounds from hops that are stable in the light and protect beer from the sunstruck reaction (ibid.). For example, rho-iso-alpha-acids, which make beer much less vulnerable to the sunstruck reaction, can be prepared by hydrogenation (reaction with sodium borohydride, NaBH$_4$) of iso-alpha-acids. Tetra-hydro-iso-alpha-acids and hexa-hydro-iso-alpha-acids can also be easily prepared from hop iso-alpha-acids. These bitter-tasting hop compounds protect beer from the sunstruck reaction. Apparently, tetra-hydro-iso-alpha-acids can be destroyed by light but do not form foul-smelling mercaptans.

Riboflavin

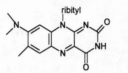

Roseoflavin

Figure A9. Riboflavin and roseoflavin (Chapter 9). *Phycomyces* is a primitive fungus that bears reproductive spores on the tip of long and slender stalks, called sporangiophores. These sporangiophores display pronounced phototropism toward ultraviolet and blue light. Based on experiments with riboflavin and roseoflavin, a flavin is implicated as the pigment that controls this phototropic reaction (M. K. Otto et al., 1981, Replacement of riboflavin by an analogue in the blue light photoreceptor of *Phycomyces*, *Proceedings of the National Academy of Sciences, USA* 78, 266–69). Riboflavin and roseoflavin have similar structures, but riboflavin absorbs blue and UV light more strongly and roseoflavin absorbs yellow light more strongly.

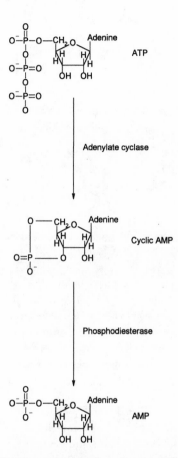

Figure A10. *Dictyostelium* amoeba migration (Chapter 10).
Dictyostelium is a slime mold that consists of independent cells (amoebas),
which aggregate into a slug and then form an erect spore-bearing structure
(P. R. Fisher, 1997, Genetics of phototaxis in a model eukaryote,
Dictyostelium discoideum, BioEssays 19, 397–407). When a population of
Dictyostelium amoebas are deprived of food, certain cells break down
adenosine triphosphate (ATP) to cyclic adenosine monophosphate
(cAMP) and then secrete this compound. In response, nearby cells migrate
toward the source of cAMP and then secrete their own cAMP.

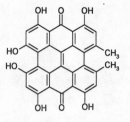

Hypericin

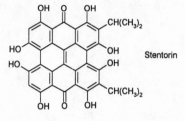

Stentorin

Figure A11. Hypericin and stentorin (Chapter 11).

Hypericin, classified by chemists as a meso-napthodianthrone-type molecule, is a naturally occurring pigment found in Saint John's wort and is a potential therapeutic agent for various diseases. Stentorin, a hypericin-like pigment, is the photoreceptive pigment that controls light-induced movement responses in the ciliated microorganism, *Stentor coeruleus;* a similar pigment, blepharismin, controls movement responses in the related species, *Blepharisma japonicum* (P. S. Song, 1995, The photo-mechanical responses in the unicellular ciliates, *Journal of Photoscience* 2, 31–35). Hypericin, stentorin, and blepharismin all strongly absorb red and ultraviolet radiation.

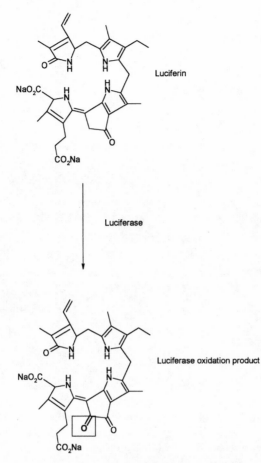

Luciferin

Luciferase

Luciferase oxidation product

Figure A12. *Gonyaulax* bioluminescence (Chapter 13).
Gonyaulax polyedra is a photosynthetic marine microorganism that emits
blue bioluminescent light at night. The compound responsible for this
bioluminescence is called *Gonyaulax* luciferin, (H. Nakamura et al., 1989,
Structure of dinoflagellate luciferin and its enzymatic and nonenzymatic
air-oxidation products, *Journal of the American Chemical Society* 111, 7607–
11). When *Gonyaulax* luciferin is oxidized by the enzyme luciferase
(note the boxed-in =O in the lower compound) it emits blue
bioluminescent light.

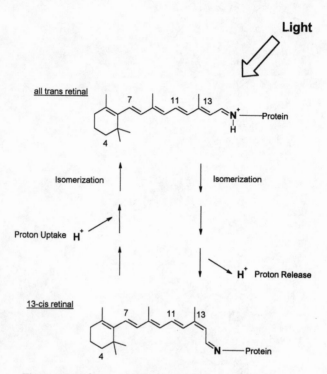

Figure A13. Photocycle of bacteriorhodopsin (Chapter 14).
In *Halobacterium,* light rapidly changes the "straight" all-trans
form of retinal in bacteriorhodopsin into the "bent" 13-cis
form of retinal, with the concomitant release of a proton
outside the cell (M. P. Krebs and G. Khorana, 1993,
Mechanism of light-dependent proton translocation by
bacteriorhodopsin, *Journal of Bacteriology* 175, 1555–60). The
13-cis form spontaneously returns to the all-trans form in a
light-independent reaction, with the concomitant uptake of a
proton from within the cell. Compare this reaction with the
photocycle of rhodopsin (see figure A1 above).

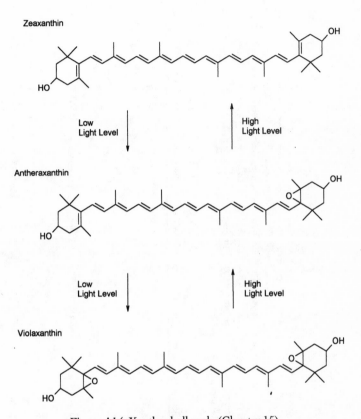

Figure A14. Xanthophyll cycle (Chapter 15).
Plants have developed several strategies for minimizing damage caused
by excess light. The xanthophyll cycle allows plants to accumulate
zeaxanthin under bright light and violaxanthin under dim light.
Zeaxanthin accepts some of the energy in light-excited chlorophyll so
this energy cannot cause photooxidative stress. (B. Demmig-Adams
and W. W. Adams, 1996, The role of xanthophyll cycle carotenoids in
the protection of photosynthesis, *Trends in Plant Science* 1, 21–26.)

Table A1

Metabolic basis and characteristic features of the porphyrias

The first column gives the biosynthetic pathway of heme, in which eight molecules of aminolevulinic acid (ALA) are transformed into one molecule of heme; second column, the enzymes that catalyze the reactions and their cellular locations (mito, mitochondria; cyto, cytoplasm); third column, the form of porphyria associated with a defect in each of the enzymes; fourth column, the main site of the enzyme defect; fifth column, the inheritance pattern (all are autosomal); sixth column, the presence of light sensitivity; and seventh column, the presence of neurological symptoms.

Compound	Enzyme	Porphyria type	Main site of defect	Inheritance pattern	Light sensitivity	Neurological symptoms
Glycine + Succinyl-CoA →	ALA synthase (mito)					
Aminolevulinic acid (ALA) →	ALA dehydratase (cyto)	Doss porphyria	Liver	Recessive	No	Yes
Porphobilinogen (PBG) →	PGB deaminase (cyto)	Acute intermittent porphyria	Liver	Dominant	No	Yes

Pathway	Enzyme	Disease	Tissue	Inheritance		
Preuroporphyrinogen →	UPG synthase (cyto)	Erythropoietic porphyria (Günther's disease)	Bone marrow	Recessive	Yes	No
Uroporphyrinogen I, III (UPG) →	UPG decarboxylase (cyto)	Porphyria cutanea tarda	Liver	Dominant	Yes	No
		Hepato erythropoietic porphyria	Liver	Recessive	Yes	No
Coproporphyrinogen I, III (CPG) →	CPG oxidase (mito)	Coproporphyria	Liver	Dominant	Yes	Yes
Protoporphyrinogen IX (PPG) →	PPG oxidase (mito)	Variegate porphyria	Liver	Dominant	Yes	Yes
Protoporphyrin IX →	Ferrochelatase (mito)	Erythropoietic protoporphyria	Bone marrow	Dominant	Yes	No
Heme						

Sources: A. Kappas et al., 1995, The porphyrias, pp. 2103–59 in The Metabolic and Molecular Bases of Inherited Disease (C. R. Scriver et al., eds.), New York, McGraw Hill, 2: 2103; M. J. Warren et al., 1996, The maddening business of King George III and porphyria, Trends in Biochemical Science 21, 229–34.

Notes

Introduction

1. A. C. Dass, 1984, *Sun-Worship in Indo-Aryan Religion and Mythology*, Delhi, Indian Book Gallery.
2. E. F. Zalewski, 1995, Radiometry and photometry, pp. 24.3–24.51 in *Handbook of Optics* (Michael Bass, ed.), New York, McGraw Hill.
3. Ibid.
4. L. O. Björn and T. C. Vogelmann, 1996, Quantifying light and ultraviolet radiation in plant biology, *Photochemistry and Photobiology* 64, 403–6.
5. K. M. Hartmann and W. Nezadal, 1990, Photocontrol of weeds without herbicides, *Naturwissenschaften* 77, 158–63.
6. J. N. Lythgoe, 1979, *The Ecology of Vision*, Oxford, Clarendon.
7. H. K. Hartline, 1959, Introduction to "Vision," pp. 615–19 in *Handbook of Physiology*, vol. 1, *Neurophysiology*. Washington, American Physiological Society. Hartline, who won the 1967 Nobel Prize in physiology or medicine with George Wald and Ragnar Granit "for their discoveries concerning the primary physiological and chemical visual processes in the eye," measured electrical responses of single optic nerve fibers in horseshoe crabs and frogs and is best known for discovery of lateral inhibition, the inhibition of sensory neural pathways by sideways connections. See nobel.sdsc.edu/laureates/medicine-1967.html
8. W. F. Kaufmann and K. M. Hartmann, 1989, Low-cost digital spectroradiometer, *Photochemistry and Photobiology* 49, 769–74.
9. W. M. Horspool and P.-S. Song (eds.), 1995, *CRC Handbook of Organic Photochemistry and Photobiology*, Boca Raton, CRC Press.
10. D. Ackerman, 1990, *A Natural History of the Senses*, New York, Random House.

1. Vision at the Threshold

1. A. Fein and E. Z. Szuts, 1982, *Photoreceptors: Their Role in Vision*, New York, Cambridge University Press; P. Buser and M. Imbert, 1995, *Vision*. Cambridge, MIT Press.

2. S. Hecht et al., 1942, Energy, quanta, and vision, *Journal of General Physiology* 25, 819–40.

3. M. H. Pirenne, 1967, *Vision and the Eye*, London, Chapman and Hall; M. H. Pirenne, 1956, Physiological mechanisms of vision and the quantum nature of light, *Biological Review* 31, 194–242.

4. Rate of light absorption: Hecht et al., 1942.

5. Buser and Imbert, 1995.

6. S. P. Langley, 1889, Energy and vision, *Philosophical Magazine* 27 (5), 1.

7. Fein and Szuts, 1982; Buser and Imbert, 1995.

8. Hecht et al., 1942.

9. M. G. F. Fourtes and S. Yeandle, 1964, Probability of occurrence of discrete potential waves in the eye of Limulus, *Journal of General Physiology* 47, 443–63.

10. D. A. Baylor et al., 1989, Responses of retinal rods to single photons, *Journal of Physiology, London* 242, 685–727. D. Baylor et al., 1984, The photocurrent, noise and spectral sensitivity of rods of the monkey, *Journal of Physiology, London* 357, 575–607.

11. T. D. Lamb, 1987, Sources of noise in photoreceptor transduction, *Journal of the Optical Society of America*, A 4, 2295–2300.

12. W. Bialek, 1987, Physical limits to sensation and perception, *Annual Review of Biophysical Chemistry* 16, 455–78.

13. D. Baylor et al., 1980, Two components of electrical dark noise in toad retinal rod outer segments, *Journal of Physiology, London* 309, 591–621. J. F. Ashmore and G. J. Falk, 1977, An analysis of voltage noise in rod bipolar cells of the dogfish retina, *Journal of Physiology, London* 332, 272–97; Baylor et al., 1984; P. G. Lillywhite, 1977, Single photon signals and transduction in an insect eye, *Journal of Comparative Physiology* 122, 189–200.

14. Baylor et al., 1984.

15. A. C. Aho et al., 1988, Low retinal noise in animals with low body temperature allows high visual sensitivity, *Nature* 34, 348–50.

16. H. B. Barlow, 1956, Retinal noise and absolute threshold, *Journal of the Optical Society of America* 46, 634–39. H. B. Barlow, 1988, The thermal limit to seeing, *Nature* 334, 296–97.

17. R. B. Barlow et al., 1993, On the molecular origin of photoreceptor noise, *Nature* 366, 64–66.

18. Ibid.

19. Ibid.

20. R. R. Birge and R. B. Barlow, 1995, On the molecular origins of thermal noise in vertebrate and invertebrate photoreceptors, *Biophysical Chemistry* 55, 115–26.

21. G. Burgula et al., 1989, Light-evoked changes in extracellular pH in frog retina, *Vision Research* 29, 1069–77; B. Oakley 2d and R. Wren, 1989, Extracellular pH in the isolated retina of the toad in darkness and during illumination, *Journal of Physiology, London* 419, 353–78.

22. Bialek, 1987.

23. S. J. Gould, 1980, *The Panda's Thumb.* New York, Norton.

2. The Five Percent Solution to Vision

1. C. Darwin, 1859, *The Origin of Species,* London, Murray.

2. W. Paley, 1807, *Natural Theology,* London, R. Faulder and Son.

3. D. Hume, 1740, *A Treatise of Human Nature,* Edinburgh, Akros.

4. R. Dawkins, 1987, *The Blind Watchmaker,* New York, Norton.

5. Darwin, 1859.

6. A. Fein and E. Z. Szuts, 1982, *Photoreceptors: Their Role in Vision,* New York, Cambridge University Press; P. Buser and M. Imbert, 1995, *Vision,* Cambridge, MIT Press.

7. S. Yokoyama and R. Yokoyama, 1996, Adaptive evolution of photoreceptors and visual pigments in vertebrates, *Annual Review of Ecology and Systematics* 27, 543–67; S. Yokoyama, 1997, Molecular genetic basis of adaptive selection: Examples from color vision in vertebrates, *Annual Review of Genetics* 31, 315–36.

8. T. Goldsmith, 1990, Optimization, constraint, and history in the evolution of eyes, *Quarterly Review of Biology* 65, 281–321.

9. P. Hegemann, 1997, Vision in microalgae, *Planta* 203, 265–74.

10. K. W. Foster et al., 1984, A rhodopsin is the functional photoreceptor for phototaxis in the unicellular eukaryote *Chlamydomonas, Nature* 311, 756–59.

11. M. A. Ali (ed.), 1984, *Photoreception and Vision in Invertebrates,* New York, Plenum.

12. L. v. Salvini-Plawen and E. Mayr, 1977, On the evolution of photoreceptors and eyes, *Evolutionary Biology* 10, 207–63.

13. M. F. Land, 1980, Optics and vision in invertebrates, pp. 471–592 in *Handbook of Sensory Physiology* (H. Autrum, ed.), Berlin, Springer-Verlag; Ali, 1984; D.-E. Nilsson, 1990, From cornea to retinal image in vertebrate eyes, *Trends in Neurosciences* 13, 55–64.

14. Nilsson, 1990.

15. Dawkins, 1987.

16. Ibid.; R. Dawkins, 1994, The eye in a twinkling, *Nature* 368, 690–91.

17. Dawkins, 1987.

18. D.-E. Nilsson and S. Pelger, 1994, A pessimistic estimate of the time required for an eye to evolve, *Proceedings of the Royal Society of London* B 256, 53–58; Dawkins, 1994.

19. C. L. Van Dover, 1989, A novel eye in "eyeless" shrimp from hydrothermal vents of the Mid-Atlantic Ridge, *Nature* 337, 458–60.

20. R. C. Lakin, 1997, Retinal anatomy of *Chorocaris chacei,* a deep-sea hydrothermal vent shrimp from the Mid-Atlantic Ridge, *Journal of Comparative Neurology* 385, 503; R. O. Kuenzler et al., 1997, Retinal anatomy of new bresiliid shrimp from the Lucky Strike and Broken Spur hydrothermal vent fields on the Mid-Atlantic Ridge, *Journal of the Marine Biological Association of the United Kingdom* 77, 707. Another small orange caridean shrimp found near hydrothermal vents, formerly called *Rimicaris aurantiaca* or *Iorania concordia,* has been shown to be the juvenile form of *Rimicaris exoculata.* T. M. Shank, 1998, Molecular systematics of shrimp *(Decapoda: Bresiliidae)* from deep-sea hydrothermal vents, part 1, Enigmatic "small orange" shrimp from the Mid-Atlantic Ridge are juvenile *Rimicaris exoculata, Molecular Marine Biology and Biotechnology* 7, 88–95.

21. J. B. Corliss et al., 1977, Submarine thermal springs on the Galapagos Rift, *Science* 203, 1073.

22. P. Rona et al., 1985, Black smokers on the Mid-Atlantic Ridge, *Eos, Transactions, American Geophysical Union* 66, 936.

23. V. Tunnicliffe, 1992, Hydrothermal vent communities of the deep sea, *American Scientist* 80, 336.

24. S. Chamberlain, 1989, video of TAG hydrothermal vents.

25. P. Rona, 1986, Black smokers, massive sulfides and vent biota at the Mid-Atlantic Ridge, *Nature* 321, 33–37. The overall reaction for this process is: $CO_2 + H_2S \rightarrow$ glycogen $+ SO_4^{-2}$. G. J. Tortora et al., 1995, *Microbiology*, New York, Benjamin-Cummings.

26. A. B. Williams and P. Rona, 1986, Two new caridean shrimps *(Bresiliidae)* from a hydrothermal field on the Mid-Atlantic Ridge, *Journal of Crustacean Biology* 6, 446–62.

27. Chamberlain, 1989; A. V. Gegruk et al., 1993, Feeding specialization of bresiliid shrimps in the TAG site hydrothermal vent community, *Marine Ecology Progress Series* 98, 247; M. F. Polz and C. M. Cavanaugh, 1995, Dominance of one bacterial phylotype at a Mid-Atlantic Ridge hydrothermal vent, *Proceedings of the National Academy of Sciences, USA* 92, 7232–36; B. Casanova et al., 1993, Bacterian epibiosis impact on the morphology of shrimps associated with hydrothermalism in the Mid-Atlantic, *Cahiers de Biologie Marine* 34, 573–88.

28. Tunnicliffe, 1992.

29. P. J. O'Neill et al., 1995, The morphology of the dorsal eye of the hydrothermal vent shrimp, *Visual Neuroscience* 12, 861–75.

30. D. J. Nuckley et al., 1995, Retinal anatomy of a new species of bresiliid shrimp from a hydrothermal vent field on the Mid-Atlantic Ridge, *Biological Bulletin, Marine Biological Laboratory, Woods Hole* 190, 98.

31. Van Dover et al., 1989.

32. O'Neill et al., 1995; Nuckley et al., 1995.

33. O'Neill et al., 1995.

34. Ibid.

35. Buser and Imbert, 1995.

36. Van Dover et al., 1989.

37. D. G. Pelli and S. C. Chamberlain, 1989, The visibility of 350 C blackbody radiation by the shrimp *Rimicaris exoculata* and man, *Nature* 337, 460; C. L. Van Dover, 1994, Light at deep sea hydrothermal vents, *Eos, Transactions, American Geophysical Union* 75, 44–45; C. L. Van Dover, 1988/1989, Do "eyeless" shrimp see the light of glowing deep-sea vents? *Oceanus* 31, 47.

38. P. J. Herring et al., 1999, Are vent shrimps blinded by science? *Nature* 398, 116.

39. Pelli and Chamberlain, 1989.

40. G. Wyszecki and W. S. Stiles, 1982, *Color Science*, New York, Wiley.

41. Van Dover et al., 1989.

42. J. Flügge, 1985, *Studienbuch zur technischen Optik*, Göttingen (Uni-Taschenbuch 109), Vanderhöck and Ruprecht.

43. Darwin, 1859.

3. A More Delightful Vision

1. A. Fein and E. Z. Szuts, 1982, *Photoreceptors: Their Role in Vision*, New York, Cambridge University Press; P. Buser and M. Imbert, 1995, *Vision*, Cambridge, MIT Press.

2. J. Nathans, 1989, The genes for color vision, *Scientific American* 260, 42–49; S. L. Merbs and J. Nathans, 1992, Absorption spectra of human cone pigments, *Nature* 356, 433–35.

3. T. W. Cronin et al., 1994, The unique visual system of the mantis shrimp, *American Scientist* 82, 356.

4. J. Nathans et al., 1992, Molecular genetics of human visual pigments, *Annual Review of Genetics* 26, 401. Achromats are also extremely sensitive to light and cannot focus on small details. On the island of Pingelap in Micronesia, the incidence of achromatopsia is about one in twelve. O. Sacks, 1997, *Island of the Colorblind*, New York, A. A. Knopf.

5. S. L. Merbs and J. Nathans, 1992, *Nature* 356, 433; J. Winderickx et al., 1992, Polymorphism in red photopigment underlies variation in color matching, *Nature*, 356, 431–33.

6. B. Lewin, 1994, On neuronal specificity and the molecular basis of perception, *Cell* 79. 935–43.

7. G. H. Jacobs, 1993, The distribution and nature of colour vision among the mammals, *Biological Reviews* 68, 413–71.

8. Ibid.; G. H. Jacobs et al., Trochromatic colour vision in New World monkeys, *Nature* 382, 156–58; G. H. Jacobs, 1996, Primate photopigments and primate color vision, *Proceedings of the National Academy of Sciences, USA* 93, 577–81.

9. J. K. Bowmaker, 1998, Evolution of colour vision in vertebrates, *Eye* 12,

541–47; S. Yokoyama and R. Yokoyama, 1996, Adaptive evolution of photoreceptors and visual pigments in vertebrates, *Annual Review of Ecology and Systematics* 27, 543–67; S. Yokoyama, 1997, Molecular genetic basis of adaptive selection: Examples from color vision in vertebrates, *Annual Review of Genetics* 31, 315–36.

10. R. L. Caldwell and H. Dingle, 1976, Stomatopods, *Scientific American* 234 (1), 80–89; N. J. Marshall et al., 1994, The six-eyed stomatopod, *Endeavour* 18, 17–26.

11. Marshall et al., 1994.

12. S. Ahyong, 1997, Phylogenetic analysis of the Stomatopoda (Malacostraca), *Journal of Crustacean Biology* 17, 695; R. B. Manning, 1995, Stomatopod Crustacea of Vietnam: The legacy of Raoul Serène, *Crustacean Research*, special no. 4, Tokyo, Carcinological Society of Japan.

13. Marshall et al., 1994; Cronin et al., 1994; Caldwell and Dingle, 1976.

14. J. N. Lythgoe, 1979, *The Ecology of Vision*, Oxford, Clarendon.

15. Cronin et al., 1994.

16. M. Burrows and G. Hoyle, 1989, Neuromuscular physiology of the strike mechanism of the mantis shrimp, *Hemisquilla*, *Journal of Experimental Zoology* 179, 379–94.

17. The Lurker's Guide to Stomatopods (www.blueboard.com/mantis/).

18. "Evolutionary arms race": R. Dawkins, 1982, *The Extended Phenotype*, New York, Oxford University Press.

19. R. L. Caldwell and H. Dingle, 1975, Ecology and evolution of agonistic behavior in stomatopods, *Naturwissenschaften* 62, 214–22; Caldwell and Dingle, 1976.

20. D.-E. Nilsson, 1990, From cornea to retinal image in vertebrate eyes, *Trends in Neurosciences* 13, 55–64.

21. N. J. Marshall, 1988, A unique colour and polarization vision system in mantis shrimps, *Nature* 333, 557–60; S. Exner, 1891, *Die Physiologie der facettirten Augen von Krebsen und Insecten*, Leipzig, Deuticke.

22. H. Schiff et al., 1986, Optics, rangefinding and neuroanatomy of the eye of a mantis shrimp *Squilla mantis* (Linnaeus) (Crustacea: Stomatopoda), *Smithsonian Contributions in Zoology* 440, 1–32.

23. Marshall, 1988.

24. M. F. Land, 1984, Crustacea, pp. 321–45 in *Photoreception and Vision in Invertebrates* (M. A. Ali, ed.), New York, Plenum.

25. T. W. Cronin and N. J. Marshall, 1989, A retina with at least ten spectral types of photoreceptors in a stomatopod crustacean, *Nature* 339, 137; N. J. Marshall et al., 1991a, The compound eyes of mantis shrimps (Crustacea, Hoplocarida, Stomatopoda), part 2, Colour pigments in the eyes of stomatopod crustaceans: Polychromatic vision by serial and lateral filtering, *Philosophical Transactions of the Royal Society of London* B 334, 57–84.

26. T. H. Goldsmith and T. W. Cronin, 1993, The retinoids of seven species of mantis shrimp, *Visual Neuroscience* 10, 915–20.

27. Cronin and Marshall, 1989; Marshall et al., 1991a.

28. Marshall et al., 1991a.

29. N. J. Marshall et al., 1991b, The compound eyes of mantis shrimps (Crustacea, Hoplocarida, Stomatopoda), part 1, Compound eye structure: The detection of polarized light, *Philosophical Transactions of the Royal Society of London* B 334, 33–56.

30. C. W. Hawryshyn, 1992, Polarization vision in fish, *American Scientist* 80, 164–75; N. Shashar, et al., 1996, Polarization vision in cuttlefish: A concealed communication channel? *Journal of Experimental Biology* 199, 2077–84; T. Labhart and E. P. Meyer, 1999, Detectors for polarized skylight in insects: A survey of ommatidial specializations in the dorsal rim area of the compound eye, *Microscopic Research Techniques* 47, 368–79; R. Wehner, 1989, Neurobiology of polarization vision, *Trends in Neuroscience* 12, 353–59; Fein and Szuts, 1982.

31. T. H. Waterman, 1981, Polarization sensitivity, pp. 281–469 in *Handbook of Sensory Physiology* (vol. VII/6C) (H. Autrum, ed.), New York, Springer-Verlag.

32. Marshall et al., 1991b.

33. Ibid.

34. M. F. Land et al., 1990, The eye-movements of the mantis shrimp Odontodactylus scyllarus (Crustacea: Stomatopoda), *Journal of Comparative Physiology* A 167, 155–66.

35. T. W. Cronin, 1986, Optical design and evolutionary adaptation in crustacean compound eyes, *Journal of Crustacean Biology* 6, 1–23.

36. J. Jones, 1994, Architecture and composition of the muscles that drive stomatopod eye movements, *Journal of Experimental Biology* 188, 317–31.

37. T. W. Cronin et al., 1988, Visual tracking of rapidly moving targets by stomatopod crustaceans, *Journal of Experimental Biology* 138, 155–79.
38. Jones, 1994; Cronin et al., 1988; T. W. Cronin et al., 1992, Regional specialization for control of ocular movements in the compound eyes of a stomatopod crustacean, *Journal of Experimental Biology* 171, 373–93.
39. Cronin et al., 1988.

Notes to Sidebars

° R. Barnes, 1990, *Monet by Monet*, New York, Knopf. A cataract is a clouding of the lens that interferes with normal vision and is an extreme condition of normal aging of the lens. Cataracts can be caused by excessive exposure to ultraviolet radiation.

† R. W. Young, 1991, *Age-Related Cataracts*, New York, Oxford University Press.

‡ J. S. Werner, 1998, Aging through the eyes of Monet, in *Color Vision* (W. Backhaus et al., eds), Berlin, Walter de Gruyter.

§ The painting he made with his right eye open was dominated by orange, yellow, and red; the painting he made with his left eye open was dominated by blue. Ibid.

Ibid.

°° S. Ahyong, 1997, Phylogenetic analysis of the Stomatopoda (Malacostraca), *Journal of Crustacean Biology* 17, 695; R. B. Manning, 1995, Stomatopod Crustacea of Vietnam: The legacy of Raoul Serène, *Crustacean Research*, special no. 4, Tokyo, Carcinological Society of Japan.

†† A. Fein and E. Z. Szuts, 1982, *Photoreceptors: Their Role in Vision*, New York, Cambridge University Press.

‡‡ M. G. J. Minnaert, 1974, *Light and Color in the Outdoors* (trans. and rev. L. Seymour, 1993), New York, Springer-Verlag; J. Strong, 1958, *Concepts of Classical Optics*, San Francisco, W. H. Freeman; D. M. Summers et al., 1970, Physical model for Haidinger's brush, *Journal of the Optical Society of America* 60, 271–72.

§§ G. P. Misson, 1993, Form and behaviour of Haidinger's brushes, *Ophthalmic and Physiological Optics* 13, 392.

4. A Burning Issue

1. F. S. Rowland, 1991, Stratospheric ozone depletion, *Annual Review of Physical Chemistry* 42, 731–68.
2. F. R. de Gruijl, 1995, Impacts of a projected depletion of the ozone layer, *Consequences* 1 (2), 13–21.
3. Rowland, 1991.
4. D.-P. Hader and M. Tevini, 1987, *General Photobiology*, New York, Pergamon.
5. Rowland, 1991.
6. S. Madronich et al., 1998, Changes in biologically active radiation reaching the earth's surface, *Journal of Photochemistry and Photobiology* B 46, 5–19.
7. M. McFarland and J. Kaye, 1992, Chlorofluorocarbons and ozone, *Photochemistry and Photobiology* 55, 911–29.
8. Madronich et al., 1998; Rowland, 1991.
9. M. J. Molina and F. S. Rowland, 1974, Stratospheric sink for chlorofluoromethanes: Chlorine atom-catalyzed destruction of ozone, *Nature* 249, 810–12.
10. Rowland, 1991.
11. Madronich et al., 1998.
12. Nobel Foundation (nobel.sdsc.edu/).
13. J. C. Farman et al., 1985, Large losses of total ozone in Antarctica reveal seasonal ClO_X/NO_X interaction, *Nature* 315, 207–10.
14. Madronich et al., 1998; M. R. Schoeberl et al., 1990, Stratospheric constituent trends from ER-2 profile data, *Geophysical Research Letters* 17, 469–72.
15. Rowland, 1991; Madronich et al., 1998.
16. McFarland and Kaye, 1992; Madronich et al., 1998.
17. J. Urbach, 1997, Ultraviolet radiation and skin cancer of humans, *Journal of Photochemistry and Photobiology* B 40, 3–7; Madronich et al., 1998.
18. Madronich et al., 1998.
19. Rowland, 1991.
20. Turco et al., 1990.
21. C. Brühl and P. Crutzen, 1989, On the disproportionate role of tropo-

spheric ozone as a filter against solar UV-B radiation, *Geophysical Research Letters* 16, 703–6.

22. P. J. Barnes, 1995, Air pollution and asthma: molecular mechanisms, *Molecular Medicine Today* 1, 149–55.

23. De Gruijl, 1995; A. B. Britt, 1996, DNA damage and repair in plants, *Annual Review of Plant Physiology* 47, 75–100.

24. De Gruijl, 1995.

25. A. R. Blaustein and D. B. Wake, 1995, The puzzle of declining amphibian populations, *Scientific American,* April, 52–57; A. R. Blaustein et al., 1994, UV repair and resistance to solar UV-B in amphibian eggs: A link to population declines? *Proceedings of the National Academy of Sciences,* USA 91, 1791–95; A. R. Blaustein, 1994, Amphibians in a bad light, *Natural History,* October, 32–39.

26. D. S. Bigelow et al., 1998, The USDA radiation monitoring program, *Bulletin of the American Meteorological Society* 79, 601–15.

27. The USDA UVB Radiation Monitoring Program (uvb.nrel.colostate. edu/).

28. H. F. Holick, 1995, Environmental factors that influence the cutaneous production of vitamin D, *American Journal of Clinical Nutrition* 61, 634S–638S.

29. P. Wilton, 1995, Cod-liver oil, vitamin D and the fight against rickets, *Canadian Medical Association Journal* 152, 1516–17.

30. R. D. Utiger, 1998, The need for more vitamin D, *New England Journal of Medicine* 338, 828–29; M. K. Thomas et al., 1998, Hypovitaminosis D in medical inpatients, *New England Journal of Medicine* 338, 777–83.

31. R. W. Young, 1991, *Age-Related Cataracts,* New York, Oxford University Press.

32. A. Das, 1999, Prevention of visual loss in older adults, *Clinical and Geriatric Medicine* 15, 131–44.

33. J. C. van der Leun, 1996, UV radiation from sunlight: Summary, conclusions and recommendations, *Journal of Photochemistry and Photobiology* B 35, 237–44.

34. B. N. Ames et al., 1995, The causes and prevention of cancer, *Proceedings of the National Academy of Sciences, USA* 92, 5258–65.

35. J. Longstreth et al., 1998, Health risks, *Journal of Photochemistry and Photobiology* B 46, 20–39.

36. P. G. Lang, 1998, Malignant melanoma, *Medical Clinics of North America* 82, 1325–58; B. A. Gilchrest et al., 1999, The pathogenesis of melanoma induced by ultraviolet radiation, *New England Journal of Medicine* 340, 1341–48.

37. International Academy for Research on Cancer, 1992, *Solar and Ultraviolet Radiation*, Lyon, France, IARC.

38. A. A. Skolnick, 1991, Is ozone loss to blame for melanoma upsurge? *Journal of the American Medical Association* 265, 3218.

39. De Gruijl, 1995.

40. D. E. Brash, 1997, Sunlight and the onset of skin cancer, *Trends in Genetics* 13, 410–14; H. N. Ananthaswamy and S. Kanjilal, 1996, Oncogenes and tumor suppressor genes in photocarcinogenesis, *Photochemistry and Photobiology* 63, 428–32.

41. K. H. Kramer, 1997, Sunlight and skin cancer: another link revealed, *Proceedings of the National Academy of Sciences, USA* 94, 11–14; D. J. Leffell, 2000, The scientific basis of skin cancer, *Journal of the American Academy of Dermatology* 42, 18–22.

42. H. S. Black et al., 1997, Photocarcinogenesis: An overview, *Journal of Photochemistry and Photobiology* B 40, 29–47.

43. Ibid.; P. U. Giacomoni, 1995, Open questions in photobiology, part 2, Induction of nicks by UVA, *Journal of Photochemistry and Photobiology* B 29, 83–85. Richard Setlow and colleagues at Brookhaven National Laboratory have used a hybrid fish of the genus *Xiphophorus* to investigate the induction of melanoma by ultraviolet radiation. These hybrid fish are much more susceptible to melanoma than humans, because they lack a tumor suppressor gene and have greater expression of an oncogene. Depending on the experimental treatment, tumors occur in 5–50 percent of the ultraviolet-irradiated fish, so researchers can use a manageable number of animals for their experiments. Studies show that hybrid *Xiphophorus* are very sensitive to ultraviolet-A, with the sensitivity from 350 to 400 nm only about 100-fold less than that at 300 nm. If humans have the same wavelength sensitivity, then about 90–95 percent of human melanomas are caused by natural ultraviolet-A radiation. R. B. Setlow et al., 1989, Animal model for ultraviolet radiation-induced melanoma: Platyfish-swordtail hybrid, *Proceedings of the National Academy of Sciences, USA* 86, 8922–26; R. B. Setlow, 1993, Wavelengths effective

in induction of malignant melanoma, *Proceedings of the National Academy of Sciences, USA* 90, 6666–70; R. S. Nairn et al., 1996, Nonmammalian models for sunlight carcinogenesis: Genetic analysis of melanoma formation in *Xiphophorus* hybrid fish, *Photochemistry and Photobiology* 64, 440–48.

44. B. J. Vermeer and M. Hurks, 1994, The clinical relevance of immunosuppression by UV irradiation, *Journal of Photochemistry and Photobiology* B 24, 149–54; J. Garssen et al., 1998, Estimation of the effect of increasing UVB exposure on the human immune system and related resistance to infectious diseases and tumours, *Journal of Photochemistry and Photobiology* B 42, 167–97; T. Mohammad, 1999, Urocanic acid photochemistry and photobiology, *Photochemistry and Photobiology* B 69, 115–35.

45. D. R. English et al., 1997, Sunlight and cancer, *Cancer Causes and Control* 8, 271–83.

46. S. R. Rajski et al., 2000, DNA repair: Models for damage and mismatch recognition, *Mutation Research* 447, 49–72.

47. A. C. Halpern and J. F. Altman, 1999, Genetic predisposition to skin cancer, *Current Opinion in Oncology* 11, 132–38.

48. H. W. Park et al., 1995, Crystal structure of DNA photolyase from *Escherichia coli, Science* 268, 1866–72.

49. Class I CPD photolyase occurs in microbes, and class II CPD photolyase occurs in higher organisms. Although these enzymes appear to have similar reaction mechanisms, they have only about 20 percent DNA sequence similarity, suggesting extensive evolutionary divergence. A. Sancar, 1996, No "End of History" for photolyases, *Science* 272, 48–49.

50. S. Y. Wang (ed.), 1976, *Photochemistry and Photobiology of Nucleic Acids,* New York, Academic Press.

51. T. Todo et al., 1993, A new photoreactivating enzyme that specifically repairs ultraviolet light-induced (6-4) photoproducts, *Nature* 361, 371–74.

52. Wang, 1976.

53. Y. F. Li et al., 1993, Evidence for lack of DNA photoreactivating enzyme in humans, *Proceedings of the National Academy of Sciences, USA* 90, 4389–93.

54. The philosophically minded might counter that "absence of evidence is not evidence of an absence."

55. B. M. Sutherland and P. V. Bennett, 1995, Human white blood cells contain cyclobutyl pyrimidine dimer photolyase, *Proceedings of the National Academy of Sciences, USA* 92, 9732–36.

56. Ibid.

57. J. Moan, 1994, UV-A radiation, melanoma induction, sunscreens, solaria and ozone reduction, *Journal of Photochemistry and Photobiology* B 24, 201–3.

58. American Academy of Dermatology (www.aad.org); M. Y. Hwang et al., 1999, JAMA patient page: Skin cancer, *Journal of the American Medical Association* 281, 676.

59. P. U. Giacomoni, 1995, Open questions in photobiology, part 3, Melanin and photoprotection, *Journal of Photochemistry and Photobiology* B 29, 87–89.

60. B. A. Gilchrest, 1998, The UV-induced SOS response: Importance to aging skin, *Journal of Dermatology* 25, 775–77; M. Yaar and B. A. Gilchrest, 1998, Aging versus photoaging: Postulated mechanisms and effectors, *Journal of Investigative Dermatology Symposium Proceedings* 3, 47–51; B. A. Gilchrest et al., 1996, Mechanisms of ultraviolet light-induced pigmentation, *Photochemistry and Photobiology* 63, 1–10.

61. American Academy of Dermatology (www.aad.org); Hwang et al., 1999.

62. S. A. Miller et al., 1998, An analysis of UVA emissions from sunlamps and the potential importance for melanoma, *Photochemistry and Photobiology* 68, 63–70; L. Roza et al., 1989, UV hazards in skin associated with the use of tanning equipment, *Journal of Photochemistry and Photobiology* B 3, 281–87.

63. J. Moan and B. Johnsen, 1994, What kind of radiation is efficient in solaria, UVA or UVB? *Journal of Photochemistry and Photobiology* B 22, 77–79.

64. Moan, 1994; Moan and Johnsen, 1994.

Notes to Sidebars

° D.-P. Hader and M. Tevini, 1987, *General Photobiology*, New York, Pergamon. Some researchers consider 290 nm the dividing point between

UV-C and UV-B; some also consider 315 nm the dividing point between UV-B and UV-A.

† Ibid.

‡ H. S. Black et al., 1997, Photocarcinogenesis: An overview, *Journal of Photochemistry and Photobiology* B 40, 29–47.

§ Ibid.

Ibid.

°° B. N. Ames, 1998, Micronutrients prevent cancer and delay aging, *Toxicology Letters* 102–3, 5; I. E. Dreosti, 1998, How best to ensure daily intake of antioxidants (from the diet and supplements) that is optimal for life span, disease, and general health, *Annals of the New York Academy of Sciences* 854, 371–77.

†† S. L. Parker et al., 1997, Cancer statistics, 1997, *Cancer Journal for Clinicians* 47, 5–27.

‡‡ B. K. Armstrong and D. R. Dallas, 1996, Cutaneous malignant melanoma, pp. 212–31 in *Cancer Epidemiology and Prevention* (D. Schottenfeld and J. F. Fraumeni, eds.), New York, Oxford University Press.

§§ Scientific American Editors, 1996, Twelve major cancers, *Scientific American* 275 (3), 126–32.

Armstrong and Dallas, 1996.

°°° Scientific American Editors, 1996.

††† H. K. Koh et al., 1993, Etiology of melanoma, *Cancer Treatment and Research* 65, 1–28.

‡‡‡ Armstrong and Dallas, 1996.

§§§ Scientific American Editors, 1996.

5. A SAD Tale

1. N. E. Rosenthal, 1993, *Winter Blues*, New York, Guilford.

2. M. Shafii and S. L. Shafii, 1990, *Biological Rhythms, Mood Disorders, Light Therapy, and the Pineal Gland*, Washington, D.C., American Psychiatric Press; E. Shorter, 1997, *A History of Psychiatry: From the Era of Asylum to the Age of Prozac*, New York, Wiley.

3. Rosenthal, 1993.

4. Ibid.

5. A. J. Lewy et al., 1982, Bright artificial light treatment of a manic-de-

pressive patient with a seasonal mood cycle, *American Journal of Psychiatry* 139, 1496–98; N. E. Rosenthal et al., 1983, Seasonal cycling in a bipolar patient, *Psychiatry Research* 8, 25–31. Lewy's previous study is A. J. Lewy et al., 1980, Light suppresses melatonin secretion in humans, *Science* 210, 1267–69.

6. N. E. Rosenthal et al., 1984, Seasonal affective disorder: A description of the syndrome and preliminary findings with light therapy, *Archives of General Psychiatry* 41, 72–80.

7. K. M. Hartmann et al., 1998, Photocontrol of germination by moon and starlight, *Zeitschrift Pflanzen Krankheit Pflanzen Schutz*, Sonderheft 16, 119–27; K. M. Hartmann, 1995, Harrowing at night is half weeded, *International Symposium on Weed and Crop Resistance to Herbicides*, Córdoba, Spain.

8. N. E. Rosenthal et al., 1989, Seasonal affective disorder and visual impairment: Two case studies, *Journal of Clinical Psychiatry* 50, 469–72.

9. Shafii and Shafii, 1990; S. Zeki, 1993, *A Vision of the Brain*, Boston, Blackwell.

10. Shafii and Shafii, 1990; S. D. Wainwright and L. K. Wainwright, 1980, Regulation of the cycle in chick pineal serotonin N-acetyltransferase activity in vitro by light, *Journal of Neurochemistry* 35, 451–57.

11. M. Max et al., 1995, Pineal opsin: A nonvisual opsin expressed in chick pineal, *Science* 267, 1502–6; T. Okano et al., 1994, Pinopsin is a chicken pineal photoreceptive molecule, *Nature* 372, 94–97.

12. Shafii and Shafii, 1990. A recent study (S. S. Campbell and P. J. Murphy, 1998, Extraocular circadian phototransduction in humans, *Science* 279, 396–99) showed that light absorbed by the popliteal region (behind the knee) can also reset the body's circadian rhythm, although an earlier study (T. A. Wehr et al., 1987, Eye versus skin phototherapy of seasonal affective disorder, *American Journal of Psychiatry* 144, 753–57) showed that light absorbed by the face, neck, arms, and legs was ineffective in treating SAD. The photoreceptor for the popliteal effect is presumably present in the blood, and may be a porphyrin-based compound (see "The Purple Disease") or a cryptochrome. Y. Miyamoto and A. Sancar, 1998, Vitamin B2-based blue-light photoreceptors in the retinohypothalamic tract as the photoactive pigments for setting the cir-

cadian clock in mammals, *Proceedings of the National Academy of Sciences, USA* 95, 6097–6102.

13. H.-W. Korf et al., 1997, Signal transduction molecules in the rat pineal organ: Ca^{2+}, pCREB, and ICER, *Naturwissenschaften* 83, 535–43.

14. Shafii and Shafii, 1990.

15. I. Balzer and R. Hardeland, 1996, Melatonin in algae and plants: Possible new roles as a phytohormone and antioxidant, *Botanica Acta* 109, 180–83.

16. Shafii and Shafii, 1990.

17. Lewy et al., 1980.

18. C. J. Bojkowski et al., 1987, Suppression of nocturnal plasma melatonin and 6-sulphatoxymelatonin by bright and dim light in man, *Hormone and Metabolic Research* 19, 437–40; J. R. Gaddy et al., 1993, Pupil size regulation of threshold of light-induced melatonin suppression, *Journal of Clinical Endocrinology and Metabolism* 77, 1398–1401; M. Laakso et al., 1993, One hour exposure to moderate illuminance (500 lux) shifts the human melatonin rhythm, *Journal of Pineal Research* 15, 21–26; I. M. McIntyre et al., 1989, Human melatonin suppression by light is intensity dependent, *Journal of Pineal Research* 6, 149–56.

19. Shafii and Shafii, 1990.

20. M. C. Blehar and N. E. Rosenthal, 1989, Seasonal affective disorders and phototherapy, *Archives in General Psychiatry* 46, 469–74.

21. D. S. Schlager, 1994, Early-morning administration of short-acting beta blockers for treatment of winter depression, *American Journal of Psychiatry* 151, 1383–85.

22. Blehar and Rosenthal, 1989.

23. S. M. Reppert and D. R. Weaver, 1995, Melatonin madness, *Cell* 83, 1059–62; F. W. Turek, 1996, Melatonin hype hard to swallow, *Nature* 379, 295–96.

24. Blehar and Rosenthal, 1989.

25. Ibid.

26. R. W. Lam et al., 1996, Effects of rapid tryptophan depletion in patients with seasonal affective disorder in remission after light therapy, *Archives in General Psychiatry* 53, 41–44; A. Neumeister et al., 1997, Effects of tryptophan depletion on drug-free patients with seasonal affective dis-

order during a stable response to bright light therapy, *Archives in General Psychiatry* 54, 133–38. One tryptophan replacement study suggests that serotonin may not have such an important role. A. Neumeister et al., 1997, Rapid tryptophan depletion in drug-free depressed patients with seasonal affective disorder, *American Journal of Psychiatry* 154, 1153–55.

27. D. Garcia-Borreguero et al., 1995, Hormonal responses to the administration of m-chlorophenylpiperazine in patients with seasonal affective disorder and controls, *Biological Psychiatry* 37, 740–49; F. M. Jacobsen et al., 1994, Behavioral responses to intravenous meta-chlorophenylpiperazine in patients with seasonal affective disorder and control subjects before and after phototherapy, *Psychiatry Research* 52, 181–97; P. J. Schwartz et al., 1997, Effects of Meta-Chlorophenylpiperazine infusions in patients with seasonal affective disorder and healthy control subjects: Diurnal responses and nocturnal regulatory mechanisms, *Psychiatry Research* 54, 375–85.

28. A. J. Lewy et al., 1988, Winter depression and the phase-shift hypothesis for bright light's therapeutic effects: History, theory, and experimental evidence, *Journal of Biological Rhythms* 3, 121–34.

29. Safii and Safii, 1990.

30. Blehar and Rosenthal, 1989.

31. Safii and Safii, 1990.

32. M. Terman, 1988, On the question of mechanism in phototherapy for seasonal affective disorder: Considerations of clinical efficacy and epidemiology, *Journal of Biological Rhythms* 3, 155–72.

33. Blehar and Rosenthal, 1989; A. Wirz-Justice et al., 1993, Light therapy in seasonal affective disorder is independent of time of day or circadian phase, *Archives in General Psychiatry* 50, 929–37.

34. Safii and Safii, 1990.

35. Effect on circadian rhythm: D. B. Boivin et al., 1996, Dose-response relationships for resetting of human circadian clock by light, *Nature* 379, 540–42.

36. Blehar and Rosenthal, 1989.

37. Harrington, A. (ed.), 1997, *The Placebo Effect: An Interdisciplinary Exploration*, Cambridge, Harvard University Press.

38. M. H. Teicher et al., 1995, The phototherapy light visor: More to it than meets the eye, *American Journal of Psychiatry* 152, 1197–1202.

Notes to Sidebars

° J. H. Kellogg, 1910, *Light Therapeutics: A Practical Manual of Phototherapy for the Student and Practitioner,* Battle Creek, Michigan, Good Health Publications.

† M. C. Blehar and N. E. Rosenthal, 1989, Seasonal affective disorders and phototherapy, *Archives in General Psychiatry* 46, 469–74; J. M. Booker and C. J. Hellekson, 1992, Prevalence of seasonal affective disorder in Alaska, *American Journal of Psychiatry* 149, 1176–82; J. M. Eagles et al., 1996, Seasonal affective disorder among psychiatric nurses in Aberdeen, *Journal of Affective Disorders* 37, 129–35.

‡ Blehar and Rosenthal, 1989.

§ Booker and Hellekson, 1992; L. N. Rosen et al., 1990, Prevalence of seasonal affective disorder at four latitudes, *Psychiatry Research* 31, 131–44.

Blehar and Rosenthal, 1989.

°° P. A. Madden et al., 1996, Seasonal changes in mood and behavior: The role of genetic factors, *Archives in General Psychiatry* 53, 47–55.

†† Eagles et al., 1996; A. L. Hegde and H. Woodson, 1996, Prevalence of seasonal changes in mood and behavior during the winter months in central Texas, *Psychiatry Research* 62, 265–71; A. Magnusson, 1996, Validation of the seasonal pattern assessment questionnaire (SPAQ), *Journal of Affective Disorders* 40, 121–29.

‡‡ S. Kasper et al., 1988, Phototherapy in subsyndromal seasonal affective disorder (S-SAD) and "diagnosed" controls, *Pharmacopsychiatry* 21, 428–29.

6. The Purple Disease

1. A. Kappas et al., 1989, The Porphyrias, pp. 2103–59 in *The Metabolic Basis of Inherited Disease* (C. L. Scriver et al., eds.), New York, McGraw Hill.

2. Porphyria can be an acquired or inherited disease. In acquired porphyria, exposure to certain chemicals disrupts heme metabolism in otherwise normal people. Exposure to lead, for example, which inhibits delta-amino levulinic acid dehydratase and ferrochelatase (see figure 5A, Appendix) can cause symptoms similar to acute intermittent por-

phyria. Polychlorinated cyclic hydrocarbons also interfere with heme metabolism and can cause acquired porphyria. Kappas et al., 1989.

3. Most researchers identify eight forms of inherited porphyria (Kappas et al., 1989), though some experts say that there are more than eight forms of inherited and acquired porphyria.

4. Hematoporphyrin: W. J. Runge, 1972, Photosensitivity in porphyria, pp. 149–162 in *Photophysiology*, vol. 7 (A. C. Geise, ed.), New York, Academic Press.

5. T. K. With, 1980, A short history of porphyrins and the porphyrias, *International Journal of Biochemistry* 11, 189–200.

6. Ibid.; Runge, 1972.

7. See With, 1980.

8. F. Meyer-Betz, 1913, Untersuchungen über die Haematoporphyrins un anderen Derivate des Blut- und Gallenfarbstoffes, *Deutsche Archiv klinisch Medizin* 112, 476–503.

9. J. D. Spikes, 1989, Photosensitization, pp. 79–110 in *The Science of Photobiology* (K. C. Smith, ed.), New York, Plenum; E. Kohen et al., 1995, *Photobiology*, San Diego, Academic Press.

10. With, 1980.

11. H. Fischer and W. Zerwick, 1924, Über Uroporphyrinogen Heptamethylester und eine neue Überfuhrung von Uro-im Koproporphyrin, *Hoppe-Seylers Zeitschrift für physiologische Chemie* 137, 242–64.

12. Based on information made available in a very detailed autopsy, Petry is believed to have suffered from erythropoietic porphyria (Günther's disease). With, 1980; Runge, 1972 (see table A1, Appendix). In other words, Petry inherited a defective gene for uroporphyrinogen synthase, and this condition caused accumulation of porphyrins in his urine and his skin.

13. With, 1980.

14. R. Lemberg and R. Legge, 1949, *Hematin Compounds and Bile Pigments*, New York, Interscience; C. Rimington, 1958, Porphyrin biosynthesis, *Review of Pure and Applied Chemistry* 8, 129–60; With, 1980.

15. A. F. McDonagh and D. M. Bissell, 1998, Porphyria and porphyrinology: The past fifteen years, *Seminars on Liver Disease* 18, 3–15; G. H. Elder, 1993, Molecular genetics of disorders of haem biosynthesis, *Journal of Clinical Pathology* 46, 977–81.

16. I. Macalpine and R. Hunter, 1966, The "insanity" of King George III: A

classic case of porphyria, *British Medical Journal* 1, 65–71. The king's four recorded incidents of mental instability were October 1788 to February 1789; February to March 1801; January to March 1804; and October 1810 to January 1820.

17. M. Guttmacher, 1941, *America's Last King: An Interpretation of the Madness of King George III*, New York, Scribner.

18. Some of the eighteenth-century reports documented blue urine in the king. Macalpine and Hunter, 1966. This can be explained by the overproduction of urinary indican and may be precipitated by the constipation associated with porphyria. W. N. Arnold, 1996, King George III's urine and indigo blue, *Lancet* 347, 1811–13.

19. I. Macalpine et al., 1968, Porphyria in the royal houses of Stuart, Hanover, and Prussia, *British Medical Journal* 1, 7–18.

20. Kappas et al., 1989.

21. Macalpine et al., 1968.

22. With, 1980.

23. Macalpine et al., 1968.

24. Warren et al., 1996, The maddening business of King George III and porphyria, *Trends in Biochemical Science* 21, 229–34.

25. G. Dean, 1968a, Royal malady, *British Medical Journal* 2, 443; G. Dean, 1968b, Letters, *Scientific American* 221 (6), 8; G. Dean, 1971. On Jansz and his descendants, G. Dean, 1971, *The Porphyrias*, Philadelphia, Pittman's Medical Publishers; G. Dean, 1957, Pursuit of a disease, *Scientific American* 196 (3), 133–42.

26. See C. E. Dent, 1968, Royal malady, *British Medical Journal* 2, 311–12.

27. I. Macalpine, 1968, Letters, *Scientific American* 221 (6), 8–9.

28. Kappas et al., 1989.

29. G. Dean, 1968b; G. Dean, 1968a; Dent, 1968.

30. I. Macalpine, 1968.

31. J. T. Hindmarsh, 1997, *Lancet* 349, 364; Dent, 1968, challenging Macalpine et al., 1968.

32. This is not as far-fetched as it may sound. A recent molecular biology study of eye tissue preserved from John Dalton, the renowned nineteenth-century English chemist, showed that he suffered from deuteranopy, the lack of a visual pigment in his retina. D. Hunt et al., 1995, The chemistry of John Dalton's color blindness, *Science* 267, 984–88.

33. J. C. G. Röhl et al., 1998, *Purple Secret: Genes, "Madness," and the Royal Houses of Europe*, London, Bantam.

34. Ibid.

35. Ibid.; Warren et al., 1996.

36. Introduction of the Charlotte mutation (in intron-8) into knockout mice would prove whether this novel mutation can cause porphyria in those animals. The researchers have considered this experiment but have ruled it out on the basis of cost. They feel that it would be easier to find another porphyric patient with a mutation similar to Charlotte's. Personal communication with Martin Warren, April 9, 1999.

37. J. C. G. Röhl et al., 1998.

38. V. H. H. Green, 1993, *The Madness of Kings: Personal Trauma and the Fate of Nations*. London, Stroud.

Notes to Sidebars

° Action spectra for the photodynamic effect of porphyria show a sharp maximum near 400 nm and a smaller peaks between 500 and 600 nm, similar to the absorption spectrum of the porphyrins. C. Rimington et al., 1967, Porphyria and photosensitivity, *Quarterly Journal of Medicine* 36, 29–57; I. A. Magnus et al., 1961, Erythropoietic protoporphyria: A new porphyria syndrome with solar urticaria due to protoporphyrinaemia, *Lancet* 2, 448. The differences in the absorption spectra of uro-, copro-, and protoporphyrin are too subtle to cause differences in the physiological action spectra for the different forms of porphyria.

† J. D. Spikes, 1989, Photosensitization, pp. 79–110 in *The Science of Photobiology* (K. C. Smith, ed.), New York, Plenum.

‡ In addition to the reaction mechanism shown here, it has been suggested that uroporphyrin can also mediate cell lysis by photodynamic reactions involving hydrogen peroxide and/or the hydroxyl radical. A. Menon et al., 1989, A comparison of the phototoxicity of protoporphyrin, coproporphyrin, and uroporphyrin using a cellular system *in vitro, Clinical Biochemistry* 22, 197–200; A. Menon et al., 1991, Role of iron in the photosensitization by uroporphyrin, *Clinical Chemistry* 202, 237–42.

§ Spikes, 1989; E. Kohen et al., 1995, *Photobiology*, San Diego, Academic Press.

J. A. Carruth, 1998, Clinical applications of photodynamic therapy, *International Journal of Clinical Practice* 52, 39–42; J. V. Moore et al., 1997, The biology of photodynamic therapy, *Physics in Medicine and Biology* 42, 913–35; J. G. Straka et al., 1990, Porphyria and porphyrin metabolism, *Annual Review of Medicine* 41, 457–69.

** University of Pennsylvania, 1998, *Oncolink* (www.oncolink.upenn.edu).

†† A. Kappas et al., 1989, The Porphyrias, pp. 2103–59 in *The Metabolic Basis of Inherited Disease* (C. L. Scriver et al., eds.), New York, McGraw Hill; P. Mustajoki et al., 1989, Heme in the treatment of porphyrias and hematological disorders, *Seminars in Hematology* 26, 1–9.

‡‡ R. Kauppinen et al., 1994, Treatment of the porphyrias, *Annals of Medicine* 26, 31–38.

§§ Kappas et al., 1989.

Kauppinen et al., 1994.

*** E. Rocchi et al., 1986, Serum ferritin in the assessment of liver iron overload and iron removal therapy in porphyria cutanea tarda, *Journal of Laboratory and Clinical Medicine* 107, 36–42.

††† Kauppinen et al., 1994.

‡‡‡ Ibid.

§§§ Kappas et al., 1989; V. A. DeLeo et al., 1976, Erythropoietic protoporphyria: Ten years experience, *American Journal of Medicine* 60, 8–12.

Kappas et al., 1989; Kauppinen et al., 1994.

7. A Novel Method of Weed Control

1. L. C. Sage, 1992, *Pigment of the Imagination: A History of Phytochrome Research*, San Diego, Academic Press.

2. K. M. Hartmann and W. Nezadal, 1990, Photocontrol of weeds without herbicides, *Naturwissenschaften* 77, 158–63.

3. D.-P. Häder and M. Tevini, 1987, *General Photobiology*, New York, Pergamon. It must be noted that Engelmann's method, though groundbreaking, lacked the quantitative rigor of modern action spectroscopy. K. M. Hartmann and I. Cohnen-Unser, 1972, Analytical action spectroscopy with living systems: Photochemical aspects and attenuance, *Berichte Deutsche Botanische Gesselschaft* 85S, 481–551.

4. D. R. Kaplan and T. J. Cook, 1996, The genius of Wilhelm Hofmeister:

The origin of causal-analytical research in plant development, *American Journal of Botany* 83, 1647–60.

5. Engelmann and subsequent researchers used the rate of oxygen production to measure the photosynthetic rate: Light + $6CO_2$ + $6H_2O$ → $C_6H_{12}O_6$ + O_2.

6. Häder and Tevini, 1987.

7. H. Ikuma, 1962, Germination of Photosensitive Lettuce Seeds, Ph.D. diss., Harvard University.

8. J. D. Bewley and M. Black, 1982, *Physiology and Biochemistry of Seeds in Relation to Germination*, vol. 2, New York, Springer-Verlag.

9. Sage, 1992.

10. Ibid.

11. Ibid.

12. Ibid.

13. Hartmann and Nezadal, 1990.

14. Sage, 1992.

15. M. W. Parker et al., 1945, Action spectrum for the photoperiodic control of floral initiation in Biloxi soybean, *Science* 102, 152–54.

16. H. A. Borthwick et al., 1952a, A reversible photoreaction controlling seed germination, *Proceedings of the National Academy of Sciences, USA* 38, 662–66.

17. H. A. Borthwick et al., 1952b, The reaction controlling floral initiation, *Proceedings of the National Academy of Sciences, USA* 38, 929–33; R. E. Kendrick and G. H. M. Kronenberg (eds.), 1994, *Photomorphogenesis in Plants*, Boston, Kluwer Academic.

18. H. A. Borthwick et al., 1954, Action spectrum of light on lettuce seed germination, *Botanical Gazette* 115, 205–25.

19. W. L. Butler et al., 1959, Detection, assay, and preliminary purification of the pigment controlling photoresponsive development in plants, *Proceedings of the National Academy of Sciences, USA* 45, 1703–8.

20. Kendrick and Kronenberg, 1994.

21. Recent evidence suggests that phytochrome may indeed be an enzyme, specifically, a protein kinase. T. D. Elich and J. Chory, 1997, Phytochrome: If it looks and smells like a histidine kinase, is it a histidine kinase? *Cell* 91, 713–16; P. H. Quail, 1997, The phytochromes: A biochemical mechanism of signaling in sight? *BioEssays* 19, 571.

22. P. H. Quail et al., 1995, Phytochromes: Photosensory perception and signal transduction, *Science* 268, 675–80.

23. Kendrick and Kronenberg, 1994.

24. D. Mandoli and W. R. Briggs, 1981, Phytochrome control of two low-irradiance responses in etiolated oat seedlings, *Plant Physiology* 67, 733–39.

25. W. J. VanDerWoude, 1985, A dimeric mechanism for the action of phytochrome: Evidence from photothermal interactions in lettuce seed germination, *Photochemistry and Photobiology* 42, 655–61.

26. Quail et al., 1995.

27. G. C. Whitelam and P. F. Devlin, 1997, Roles of different phytochromes in *Arabidopsis* photomorphogenesis, *Plant, Cell, and Environment* 20, 752–58.

28. Ibid.; Quail et al., 1995.

29. Bewley and Black, 1982.

30. J. Harper, 1977, *Population Biology of Plants*, New York, Academic Press; R. Cook, 1980, The Biology of Seeds in the Soil, pp. 3–21 in *Demography and Evolution in Plant Populations* (O. T. Solbrig, ed.), San Diego, University of California Press.

31. A. L. Scopel et al., 1991, Induction of extreme light sensitivity in buried weed seeds and its role in the perception of soil cultivation, *Plant, Cell, and Environment* 14, 501–8.

32. Bewley and Black, 1982.

33. D. Cohen, 1966, Optimizing reproduction in a randomly varying environment, *Journal of Theoretical Biology* 12, 119–29.

34. Weed Science Society of America, 1998 (ext.agn.uiuc.edu/wssa/).

35. D. Pimentel and H. Lehman, 1993, *The Pesticide Question*, New York, Chapman and Hall.

36. Pimentel and Lehman, 1993.

37. Ibid.; S. Steingraber, 1997, *Living Downstream*, New York, Addison-Wesley; J. Wargo, 1996, *Our Children's Toxic Legacy: How Science and Law Fail to Protect Us from Pesticides*, New Haven, Yale University Press.

38. R. Carson, 1962, *Silent Spring*, New York, Fawcett Crest.

39. M. B. Thomas, 1999, Ecological approaches and the development of "truly integrated" pest management, *Proceedings of the National Academy of Sciences, USA* 96, 5944–51; A. J. Burn et al. (eds.), 1987, *Integrated Pest*

Management, San Diego, Academic Press; Center for Integrated Pest Management (North Carolina State University), 1998 (ipmwww.ncsu. edu/cipm/Virtual_Center.html).

40. Pimentel and Lehman, 1993.

41. Somewhat paradoxically, these two herbicides are also on the list of herbicides approved for use by the Weed Science Society of America (ext.agn.uiuc.edu/wssa/).

42. Office of Technology Assessment, 1990, *Beneath the Bottom Line: Agricultural Approaches to Reduce Agrichemical Contamination of Ground Water,* Washington, D.C., U.S. Government Printing Office.

43. Ibid.; W. F. Ritter, 1990, Pesticide contamination of ground water in the United States: A review, *Journal of Environmental Science and Health* B25, 1–29.

44. Pimentel and Lehman, 1993.

45. C. Preston et al., 1996, Multiple mechanisms and multiple herbicide resistance in *Lolium rigidum,* pp. 117–29 in *Molecular Genetics and Evolution of Pesticide Resistance* (T. M. Brown, ed.), Washington, D.C., American Chemical Society; S. B. Powles and C. Preston, 1998, Herbicide cross resistance and multiple resistance in plants (plantprotection.org/HRAC/ mono2.html).

46. S. B. Powles et al., 1990a, Herbicide cross resistance in annual ryegrass *(Lolium rigidum):* The search for a mechanism, pp. 394–406 in *ACS Symposium Series, Managing Resistance to Agrochemicals: From Fundamental Research to Practical Strategies* (M. B. Green et al., eds.), Washington, D.C., American Chemical Society Press; S. B. Powles and J. A. M. Holtum (eds.), 1994, *Herbicide Resistance in Plants: Biology and Biochemistry,* Boca Raton, Fla., Lewis/CRC Press.

47. F. Forcella et al., 1992, Weed seedbanks of the U. S. Corn Belt: Magnitude, variation, emergence, and application, *Weed Science* 40, 636–44; Cook, 1980; Harper, 1977.

48. Hartmann and Nezadal, 1990.

49. J. Sauer and G. Struik, 1964, A possible ecological relation between soil disturbance, light flash, and seed germination, *Ecology* 45, 884.

50. This line of research was also pursued by Wesson and Waring. G. Wesson and P. F. Waring, 1969, The role of light in the germination of natu-

rally occurring populations of buried weed seeds, *Journal of Experimental Botany* 20, 402–13.

51. Hartmann and W. Nezadal, 1990.

52. Ibid.

53. K. M. Hartmann and W. Nezadal, 1989, Efficient photocontrol of weeds in crop fields, p. 65 in *Photomorphogenesis in Plants, European Symposium,* Freiburg, Germany, p. 65.

54. Hartmann and Nezadal, 1990.

55. Respectively, V. Miller, 1993, Team keeping weeds in the dark, *Research Nebraska,* September, pp. 14–15; A. L. Scopel et al., 1994, Photostimulation of seed germination during soil tillage, *New Phytologist* 126, 145–52; H. Becker, 1995, Nightmare in tilling fields: A horror for weed pests, *Agricultural Research,* December, pp. 10–11; P. K. Jensen, 1995, Effect of light environment during soil disturbance on germination and emergence pattern of weeds, *Annals of Applied Biology* 127, 561–71; J. Ascard, 1992, Soil cultivation in darkness reduced weed emergence, *Acta Horticulturae* 372, 167–77; Scopel et al., 1991.

56. Hartmann and Nezadal, 1990.

57. E. Sloane, 1965, *Diary of an Early American Boy,* New York, Ballantine.

Notes to Sidebars

° J. D. Bewley and M. Black, 1982, *Physiology and Biochemistry of Seeds in Relation to Germination,* vol. 2, New York, Springer-Verlag.

† Ibid.

‡ R. Carson, 1962, *Silent Spring,* New York, Fawcett Crest.

§ G. F. Ryan, 1970, Resistance of common groundsel to simazine and atrazine, *Weed Science* 18, 614–16.

D. Pimentel and H. Lehman, 1993, *The Pesticide Question,* New York, Chapman and Hall.

°° A. J. Burn et al., 1987, *Integrated Pest Management,* San Diego, Academic Press; Center for Integrated Pest Management (North Carolina State University), 1998 (ipmwww.ncsu.edu/cipm/Virtual_Center.html).

†† Ibid.

8. Light and Beer

1. J. Templar et al., 1995, Formation, measurement and significance of lightstruck flavor in beer: A review, *Brewers Digest,* May, pp. 18–25.

2. D. De Keukeleire, 1991, Photochemistry of beer, *The Spectrum* 4 (2), 1, 3–7.

3. Technically, the resins and aromatic oils of hops are in lupulin glands appended to bracteoles, modified leaves that are part of the hop inflorescence. S. Kenny, 1997, Hunting for the heart of the hop, *Zymurgy* 20 (4), 38–39.

4. R. Daniels, 1997, Hop physiology and chemistry, *Zymurgy* 20 (4): 40–49. Alpha acids have one prenyl group and one hydroxide attached to carbon-6 of the carbon ring, whereas beta acids have two prenyl groups attached to carbon-6 of the carbon ring (see figure A7, Appendix).

5. R. Mosher, 1997, *Zymurgy*'s guide to hops, *Zymurgy* 20 (4): pull-out poster; M. Verzele and D. De Keukeleire, 1991, *Chemistry and Analysis of Hop and Beer Bitter Acids,* New York, Elsevier.

6. Verzele and De Keukeleire, 1991.

7. F. B. Salisbury and C. W. Ross, 1992, *Plant Physiology,* Belmont, Calif., Wadsworth.

8. R. H. Whittaker and P. P. Feeny, 1971, Allelochemics: chemical interactions between species, *Science* 171, 757–70; G. Rosenthal and M. R. Berenbaum (eds.), 1992, *Herbivores: Their Interaction with Secondary Plant Metabolites,* San Diego, Academic Press.

9. J. Rozema et al., 1997, UV-B as an environmental factor in plant life: Stress and regulation, *Trends in Ecology and Evolution* 12, 22–28; B. W. Shirley et al., 1995, Analysis of *Arabidopsis* mutants deficient in flavonoid biosynthesis, *Plant Journal* 8, 659–71.

10. Some would say that lagering, or cold maturation, is a fourth stage important in the brewing of lagers. W. Kunze, 1996, *Technology Brewing and Malting,* Berlin, VLB.

11. Templar et al., 1995; D. De Keukeleire and M. Verzele, 1971, The absolute configuration of the isohumulones and the humulinic acids, *Tetrahedron* 27, 4939–45.

12. Verzele and De Keukeleire, 1991.

13. D. De Keukeleire et al., 1992, The history and analytical chemistry of

beer bitter acids, *Trends in Analytical Science* 11, 275–80; De Keukeleire, 1991.

14. Verzele and De Keukeleire, 1991.

15. About one-third are trans-iso-alpha acids, and about two-thirds are cis-iso-alpha acids, the more stable form. Verzele and De Keukeleire, 1991.

16. M. Jackson, 1993, *Beer Companion,* Philadelphia, Running Press.

17. Kunze, 1996.

18. Verzele and De Keukeleire, 1991.

19. V. Peacock, 1997, Fundamentals of Hop Chemistry, *Malting and Brewing Association of America Technical Quarterly* 34, 4–8.

20. The brewers of Lambic beers, from the Senne Valley of Belgium, are a notable exception. Brewers of these spontaneously fermented beers use aged hops, which can have very high levels of oxidized alpha and beta-acids. Jackson, 1993.

21. Templar et al., 1995.

22. Ibid.; S. M. Russell et al., 1995, Spoilage bacteria of fresh broiler chicken carcasses, *Poultry Science* 74, 2041–7; J. Tonzetich, 1977, Production and origin of oral malodor: A review of mechanisms and methods of analysis, *Journal of Periodontology* 48, 13–20; Y. Yaegaki and K. Sanada, 1992, Biochemical and clinical factors influencing oral malodor in periodontal patients, *Journal of Periodontology* 63, 783–89.

23. Templar et al., 1995; J. D. Spikes, 1981, pp. 39–83 in *Photochemical and Photobiological Reviews* (K. C. Smith, ed.), New York, Plenum.

24. Kunze, 1996; Jackson, 1993; Templar et al., 1995.

25. Assuming that one drop is 0.05 ml, then one part per billion is equivalent to 1 drop per 50 million ml (0.05×10^9). One gallon is equivalent to 3,785 ml, so the ratio is about one drop per 13,210 gallons.

26. G. M. A. Blondeel et al., 1987, The photolysis of trans-isohumulone to dehydrohumulinic acid, a key route to the development of sunstruck flavour in beer, *Journal of the Chemical Society. Perkin Transactions* 1, 2715–17.

27. Templar et al., 1995.

28. Ibid.; S. Sakuma et al., 1991, Sunstruck flavor formation in beer, *Journal of the American Society of Brewing Chemists* 49, 162–65.

29. Templar et al., 1995.

30. S. Sakuma et al., 1991.

31. Ibid.

32. G. Hofmann (ed.), 1977, *ISCO Tables,* Lincoln, Neb., Instrumentation Specialties Company; D.-P. Häder and M. Tevini, 1987, *General Photobiology,* New York, Pergamon.

33. Hofmann, 1977; Häder and Tevini, 1987.

34. Kalsec Incorporated, 3713 West Main, P.O. Box 50511, Kalamazoo, Mich. 49005-0511.

35. F. R. Sharpe and I. H. L. Ormrod, 1991, Fast isomerization of humulone by photo-reaction preparation of an HPLC standard, *Journal of the Institute of Brewing* 97, 33–38; De Keukeleire, 1991.

36. De Keukeleire, 1991.

37. Ibid.

38. J. C. Andre et al., 1988, Industrial photochemistry, part 11, Comparison between different types of photoreactors and selective filtering for monomolecular photoreactions, *Journal of Photochemistry and Photobiology* A 42, 383–96.

39. De Keukeleire, 1991.

40. Ibid.

Notes to Sidebars

° G. M. A. Blondeel et al., 1987, The photolysis of trans-isohumulone to dehydrohumulinic acid, a key route to the development of sunstruck flavour in beer, *Journal of the Chemical Society. Perkin Transactions* 1, 2715–17.

† J. Templar et al., 1995, Formation, measurement and significance of lightstruck flavor in beer: A review, *Brewers Digest,* May, pp. 18–25; S. Sakuma et al., 1991, Sunstruck flavor formation in beer, *Journal of the American Society of Brewing Chemists* 49, 162–65.

‡ D. De Keukeleire, 1993, The lightstruck flavour of beer, *EPA Newsletter* 49, 28–36.

9. Phycomyces, *the Fungus That Sees*

1. A. Fein and E. Z. Szuts, 1982, *Photoreceptors: Their Role in Vision,* New York, Cambridge University Press.

2. P. Buser and M. Imbert, 1995, *Vision*, Cambridge, MIT Press.

3. D. C. Hood, 1998, Lower-level visual processing and models of light adaptation, *Annual Review of Psychology* 49, 503–35; Buser and Imbert, 1995.

4. Human senses: H. J. Chiel and R. D. Beer, 1997, The brain has a body: Adaptive behavior emerges from interactions of nervous system, body and environment, *Trends in Neurosciences* 20, 553–57; V. Torre et al., 1995, Transduction and adaptation in sensory receptor cells, *Journal of Neuroscience* 15, 7757–68; plants: P. Galland, 1989, Photosensory adaptation in plants, *Botanica Acta* 102, 11–20; microorganisms: J. P. Armitage, 1992, Behavioral responses in bacteria, *Annual Review of Physiology* 54, 683–714.

5. See Hood, 1998; Buser and Imbert, 1995; Fein and Szuts, 1982; Galland, 1989.

6. *Phycomyces blakesleeanus* was named after Albert Francis Blakeslee. Blakeslee discovered sexuality and the presence of different mating types in the fungi while performing research on bread molds as a graduate student. In acknowledgment of this fundamental discovery, *Phycomyces blakesleeanus* was named in his honor. J. F. Crow, 1997, Birth defects, Jimson weeds and bell curves, *Genetics* 147, 1–6.

7. E. Cerda'-Olmedo, 1987a, A Biography of Phycomyces, pp. 7–26 in *Phycomyces* (E. Cerda'-Olmedo and E. D. Lipson, eds.), Cold Spring Harbor, N.Y., Cold Spring Harbor Laboratory.

8. E. Cerda'-Olmedo, 1987b, Standard Growth Conditions and Variations, pp. 337–39 in *Phycomyces* (Cerda'-Olmedo and Lipson, eds.).

9. Cerda'-Olmedo, 1987a.

10. Ibid.

11. P. Galland and E. D. Lipson, 1987, Light Physiology of *Phycomyces* sporangiophores, pp. 49–92 in: *Phycomyces* (Cerda'-Olmedo and Lipson, eds.). On Hofmeister, see D. R. Kaplan and T. J. Cook, 1996, The genius of Wilhelm Hofmeister: The origin of causal-analytical research in plant development, *American Journal of Botany* 83, 1647–60.

12. C. Darwin, 1880, *The Power of Movement in Plants*, London, Murray.

13. Galland and Lipson, 1987; W. Shropshire Jr., 1987, Bibliography, pp. 381–419 in *Phycomyces* (Cerda'-Olmedo and Lipson, eds.).

14. E. S. Castle, 1940, Discontinuous growth of single plant cells measured

at short intervals and the theory of intussusception, *Journal of Cellular and Comparative Physiology* 15, 285–98.

15. P. A. Ensminger and E. D. Lipson, 1992, Growth rate fluctuations in *Phycomyces* sporangiophores, *Plant Physiology* 99, 1376–80; P. A. Ensminger and M. Vinson, 1994, Deterministic non-linear dynamics in *Phycomyces* sporangiophores, *Journal of Theoretical Biology* 170, 259–66.

16. Ensminger and Vinson, 1994.

17. Ensminger and Lipson, 1992.

18. H. F. Judson, 1979, *The Eighth Day of Creation*, New York, Simon and Schuster; E. P. Fischer and C. Lipson, 1988, *Thinking About Science*, New York, Norton; E. P. Fischer, 1988, *Das Atom der Biologen*, Munich, Piper.

19. J. D. Watson and F. H. C. Crick, 1953, A structure for deoxyribose nucleic acid, *Nature* 171, 737–38.

20. Personal communication from David S. Dennison, Feb. 21, 1995.

21. M. Delbrück and W. Reichardt, 1956, System analysis for the light growth reactions in *Phycomyces*, pp. 3–44 in *Cellular Mechanisms in Differentiation and Growth* (D. Rudnick, ed.), Princeton, Princeton University Press.

22. Fein and Szuts, 1982; Buser and Imbert, 1995.

23. K. Bergman et al., 1969, *Phycomyces*, *Bacteriological Reviews* 33, 99–150.

24. M. Delbrück, 1976, Light and life, *Carlsberg Laboratory Research Communications* 41, 299–309; E. D. Lipson and B. A. Horwitz, 1991, Photosensory Reception and Transduction, pp. 1–64 in *Sensory Receptors and Signal Transduction* (J. L. Spudich and B. H. Satir, eds.), New York, Wiley.

25. D. Presti and P. Galland, 1987, Photoreceptor Biology of Phycomyces, pp. 93–126 in *Phycomyces* (Cerda'-Olmedo and Lipson, eds.).

26. M. K. Otto et al., 1981, Replacement of riboflavin by an analogue in the blue light photoreceptor of *Phycomyces*, *Proceedings of the National Academy of Sciences, USA* 78, 266.

27. G. Löser and E. Schäfer, 1980, Phototropism in *Phycomyces*: A photochromic sensor pigment? pp. 43–57 in *The Blue Light Syndrome* (H. Senger, ed.), New York, Springer-Verlag; G. Löser and E. Schäfer, 1984, Photogeotropism of *Phycomyces*: Evidence for more than one photoreceptor, pp. 118–24 in: *Blue Light Effects in Biological Systems* (H. Senger, ed.), New York, Springer-Verlag; G. Löser and E. Schäfer, 1986,

Are there several photoreceptors involved in phototropism of *Phy-comyces blakesleeanus*? Kinetic studies of dichromatic irradiation, *Photo-chemistry and Photobiology* 43, 195. Strictly speaking, this is true only for oxidized flavin, the form I am referring to in the main text. Reduced or semireduced flavins may absorb these longer wavelengths and have been suggested as possible photoreceptive pigments. P. Galland et al., 1989a, Subliminal light control of dark adaptation kinetics in *Phycomyces* phototropism, *Photochemistry and Photobiology* 49, 485–91; P. Galland et al., 1989b, Light-controlled adaptation kinetics in *Phycomyces:* Evidence for a novel yellow-light absorbing pigment, *Photochemistry and Photobiology* 49, 493–99.

28. Galland at al., 1989a; Galland et al., 1989b.
29. Ibid.
30. P. Galland and H. Senger, 1988, The role of flavins as photoreceptors, *Journal of Photochemistry and Photobiology* B 1, 277–94.
31. Galland et al., 1989; X.-Y. Chen et al., 1993, Action spectrum for sub-liminal light control of adaptation in *Phycomyces* phototropism, *Photo-chemistry and Photobiology* 58, 425–31.
32. Chen et al., 1993.

Notes to Sidebars

° E. S. Castle, 1940, Discontinuous growth of single plant cells measured at short intervals and the theory of intussusception, *Journal of Cellular and Comparative Physiology* 15, 285–98.

† P. Galland and E. D. Lipson, 1987, Light Physiology of *Phycomyces* spo-rangiophores, pp. 49–92 in *Phycomyces* (E. Cerda'-Olmedo and E. D. Lipson, eds.), Cold Spring Harbor, N.Y., Cold Spring Harbor Laboratory.

‡ Level of light (energy flux) that causes a "just-detectable" bending re-sponse.

10. Dictyostelium, *the Amoeba and the Slug*

1. O. Brefeld, 1869, *Dictyostelium mucoroides:* Ein neuer Organismus aus der Verwandschaft der Myxomyceten, *Abhandlung senckenberg Naturforschung*

Gesellschaft Frankfurt 7, 87–107; K. B. Raper, 1984, *The Dictyostelids*, Princeton, Princeton University Press.

2. L. Margulis and K. V. Schwartz, 1998, *Five Kingdoms*, New York, Freeman.

3. C. J. Alexopoulis et al., 1996, *Introductory Mycology*, New York, Wiley; T. D. Bruns et al., 1991, Fungal molecular systematics, *Annual Review of Ecology and Systematics* 22, 525–64; L. S. Olive, 1975, *The Mycetozoans*, New York, Academic Press; T. D. Bruns, et al., 1993, Evolutionary relationships within the fungi: analyses of nuclear small subunit rRNA sequences, *Molecular Phylogenetics and Evolution* 1, 231–41; L. S. Olive, 1969, Letters, *Science* 164, 157; R. H. Whitaker, 1969, Letters, *Science* 164, 157.

4. Alexopoulos et al., 1996.

5. K. B. Raper, 1935, *Dictyostelium discoideum*, a new species of slime mold from decaying forest leaves, *Journal of Agricultural Research* 50, 135–47.

6. J. T. Bonner, 1974, *On Development*, Cambridge, Harvard University Press.

7. Alexopoulis et al., 1996. See J. T. Bonner, 1959, *The Cellular Slime Molds*, Princeton, Princeton University Press.

8. J. Franke, 1997, The Franke Dictyostelium Literature Database (dicty.cmb.nwu.edu/dicty/reference_database/index.html).

9. P. J. M. Van Haastert et al., 1991, Sensory transduction in eukaryotes: A comparison between *Dictyostelium* and vertebrate cells, *European Journal of Biochemistry* 195, 289–303; J. Van Houten, 1994, Chemosensory transduction in eukaryotic microorganisms: Trends for neuroscience? *Trends in Neurosciences* 17, 62–71; A. A. Noegel and J. E. Luna, 1995, The *Dictyostelium* cytoskeleton, *Experientia* 51, 1135–43.

10. *Dictyostelium* amoebas are more technically referred to as "myxamoebae." Alexopoulis et al., 1996. Some *Dictyostelium* papers have used units of lux for quantifying levels of light. Lux is a photometric unit that is based on the particular characteristics of the human eye and should not be applied to *Dictyostelium*, which clearly does not have eyes and does not use the pigment rhodopsin for sensing light. For papers cited below that have used units of lux, I used the conversions of $1 \ W \ m^{-2} = 29.7$ lux for white light from a tungsten filament lamp and $1 \ W \ m^{-2} = 302$ lux for white light from a halogen lamp. D.-P. Häder, 1986, *General Photobiology*, New York, Pergamon.

11. P. Pan et al., 1972, Folic acid as a second chemotactic substance in the cellular slime molds, *Nature* 237, 181–82.

12. The threshold values for amoebas quoted here were determined for strain AX-2. D.-P. Häder and B. Vollersten, 1991, Phototactic orientation in *Dictyostelium discoideum* amoebae, *Acta Protozoologica* 30, 19–24.

13. Häder and Vollertsen, 1991.

14. D.-P. Häder et al., 1983, Responses of *Dictyostelium discoideum* amoebae to local stimulation by light, *Cell Biology International Reports* 7, 611–15.

15. K. M. Hartmann, 1983, Action spectroscopy, pp. 542–73 in *Biophysics* (W. Hoppe et al., eds.), New York, Springer-Verlag; K. M. Hartmann and I. Cohnen-Unser, 1972, *Berichte Deutsche Botanische Gesselschaft* 85, 481. For action spectrum research, see D.-P. Häder and K. L. Poff, 1979a, Light-induced accumulation of *Dictyostelium discoideum* amoebae, *Photochemistry and Photobiology* 29, 1157–61; D.-P. Häder and K. L. Poff, 1979b, Photodispersal from light traps by amoebas of *Dictyostelium discoideum*, *Experimental Mycology* 3, 121–31; D.-P. Häder et al., 1988, Multiple photoreceptors in phototaxis of *Dictyostelium discoideum* amoebae, *Protoplasma* 1, 155–61.

16. Häder et al., 1988.

17. T. Schlenkrich et al., 1995, Biochemical and spectroscopic characterization of the putative photoreceptor for phototaxis in amoebae of the cellular slime mould, *Journal of Photochemistry and Photobiology* B 30, 139–43; H.-P. Vornlocher and D.-P. Häder, 1992, Isolation and characterization of the putative photoreceptor for phototaxis in amoebae of the cellular slime mold, *Botanica Acta* 105, 47–54.

18. W. A. Pryor, 1976, *Free Radicals in Biology*, San Diego, Academic Press; J. D. Spikes, 1989, Photosensitization, pp. 79–110 in *The Science of Photobiology* (K. C. Smith, ed.), New York, Plenum.

19. A. Kappas et al., 1989, The Porphyrias, pp. 2103–59 in *The Metabolic Basis of Inherited Disease* (C. L. Scriver et al., eds.), New York, McGraw Hill.

20. A. U. Khan and T. Wilson, Reactive oxygen species as cellular messengers, *Current Biology* 2, 437–45.

21. R. H. Kessin et al., 1992, The development of a social amoeba, *American Scientist* 80, 556–65; J. Van Houten, 1994.

22. Ibid.

23. This description of cell migration is greatly oversimplified. C. van Oss and colleagues provide a more complete description of the dynamics of cell migration. C. van Oss et al., 1996, Spatial pattern formation during aggregation of the slime mould *Dictyostelium discoideum, Journal of Theoretical Biology* 181, 203–13.

24. This mass is called a pseudoplasmodium because, unlike the true slime molds (Myxomycetes), which form a true plasmodium, the constituent cells retain their cell membranes. Alexopoulis et al., 1996.

25. P. R. Fisher, 1997, Genetics of phototaxis in a model eukaryote, *Dictyostelium discoideum, BioEssays* 19, 397–407.

26. D. Dormann et al., 1997, Twisted scroll waves organize *Dictyostelium mucoroides* slugs, *Journal of Cell Science* 110, 1831–37; F. Siegert and C. J. Weijer, 1992, Three-dimensional scroll waves organize *Dictyostelium* slugs, *Proceedings of the National Academy of Sciences, USA* 89, 6433–37.

27. P. R. Fisher, 1997. Ammonia clearly has a role in barrier avoidance (J. T. Bonner et al., 1988, The possible role of ammonia in phototaxis of migrating slugs of *Dictyostelium discoideum, Proceedings of the National Academy of Sciences, USA* 85, 3885–87; J. T. Bonner, 1989, Ammonia and thermotaxis: further evidence for a central role of ammonia in the directed cell mass movements of *Dictyostelium discoideum, Proceedings of the National Academy of Sciences, USA* 86, 2733–36), but Paul R. Fisher has questioned its role in phototaxis and thermotaxis. P. R. Fisher, 1991, The role of gaseous metabolites in phototaxis by *Dictyostelium discoideum* slugs, *FEMS Microbiology Letters* 77, 117–20.

28. D. W. Francis, 1964, Some studies on phototaxis of *Dictyostelium, Journal of Cellular and Comparative Physiology* 64, 131–38; D.-P. Häder and U. Burkhart, 1983, Optical properties of *Dictyostelium discoideum* pseudoplasmodia responsible for phototactic orientation, *Experimental Mycology* 7, 1–8.

29. D.-P. Häder, 1985, Negative phototaxis of *Dictyostelium discoideum* pseudoplasmodia in UV radiation, *Photochemistry and Photobiology* 41, 225–28.

30. P. R. Fisher et al., 1981, An extracellular chemical signal controlling phototactic behavior by *Dictyostelium discoideum* slugs, *Cell* 23, 799–807; Fisher, 1997.

31. This threshold was determined for the NC4 strain. K. L. Poff and D.-P.

Häder, 1984, An action spectrum for phototaxis by pseudoplasmodia of *Dictyostelium discoideum*, *Photochemistry and Photobiology* 39, 433–36. The threshold for the AX2 strain is about 100-fold lower. D.-P. Häder and Haser, 1991, *Botanica Acta* 104, 200.

32. D. W. Francis, 1964; Poff and Häder, 1984; K. L. Poff et al., 1973, Light-induced absorbance changes associated with phototaxis in *Dictyostelium*, *Proceedings of the National Academy of Sciences, USA* 70, 813–16.

33. D.-P. Häder et al., 1980, Responses to light by a nonphototactic mutant of *Dictyostelium discoideum*, *Experimental Mycology* 4, 382–85.

34. High–molecular weight heme protein: Poff et al., 1973; flavin-containing protein and cytochrome: K. L. Poff and W. L. Butler, 1974, Spectral characteristics of the photoreceptor pigment of phototaxis in *Dictyostelium discoideum*, *Photochemistry and Photobiology* 20, 241–44.

35. W. L. Butler et al., 1959, Detection, assay, and preliminary purification of the pigment controlling photoresponsive development in plants, *Proceedings of the National Academy of Sciences, USA* 45, 1703–7.

36. P. K. Darcy et al., 1994, Genetic analysis of *Dictyostelium* slug phototaxis mutants, *Genetics* 137, 977–85.

37. P. K. Darcy et al., 1993, Phototaxis genes on linkage group V in *Dictyostelium discoideum*, *FEMS Microbiology Letters* 11, 123–28; Darcy et al., 1994.

38. Z. Wilczynska and P. R. Fisher, 1994, Analysis of a complex plasmid insertion in a phototaxis-deficient transformant of a *Dictyostelium discoideum* selected on a *Micrococcus luteus* lawn, *Plasmid* 32, 182–94.

39. D.-P. Häder and K. L. Poff, 1979, Inhibition of aggregation by light in the cellular slime mold *Dictyostelium discoideum*, *Archives of Microbiology* 123, 281–85; T. M. Konijin and K. B. Raper, 1966, *Biological Bulletin* 131, 446; K. B. Raper, 1940, Pseudoplasmodium formation and organisation in *Dictyostelium discoideum*, *Journal of the Elisha Mitchell Science Society* 56, 241–82.

40. Fisher, 1997. In nature, the *Dictyostelium* aggregate presumably may consist of genetically heterogeneous cells.

41. Bonner, 1974. Numerous researchers have investigated the "altruistic" behavior of *Dictyostelium* cells. See D. P. Armstrong, 1984, Why don't cellular slime molds cheat? *Journal of Theoretical Biology* 109, 271–83; D. Atzmony et al., 1997, Altruistic behaviour in *Dictyostelium dis-*

coideum explained on the basis of individual selection, *Current Science* 72, 142–45.

42. K. B. Raper, 1984, *The Dictyostelids*, Princeton, Princeton University Press.

43. Two key features distinguish the *Dictyostelium* sorocarp and the *Phycomyces* sporangiophore: (a) the *Phycomyces* sporangiophore is a single giant cell, whereas the *Dictyostelium* sorocarp is composed of thousands of separate cells; (b) the *Phycomyces* sporangiophore grows to more than one hundred times the length of the *Dictyostelium* sorocarp. Alexopoulis et al., 1996.

44. Paul R. Fisher, 1998, personal communication.

Notes to Sidebars

° H. C. Berg, 1993, *Random Walks in Biology*, Princeton, Princeton University Press.

† H. Yang et al., 1995, Phototaxis away from blue light by an *Escherichia coli* mutant accumulating protoporphyrin IX, *Proceedings of the National Academy of Sciences, USA* 92, 7332–36.

‡ Ibid.; J. A. Armitage, 1992, Behavioral responses in bacteria, *Annual Review of Physiology* 54, 683–714.

§ R. H. Kessin et al., 1992, The development of a social amoeba, *American Scientist* 80, 556–65.

11. High Hopes for Hypericin

1. D. Meruelo et al., 1988, Therapeutic agents with dramatic antiretroviral activity and little toxicity at effective doses: Aromatic polycyclic diones hypericin and pseudohypericin, *Proceedings of the National Academy of Sciences, USA* 85, 5230–34; G. Lavie et al., 1989, Studies of the mechanisms of action of the antiretroviral agents hypericin and pseudohypericin, *Proceedings of the National Academy of Sciences, USA* 86, 5963–67.

2. Equine infectious anemia virus: S. Carpenter and G. A. Kraus, 1991, Photosensitization is required for inactivation of equine infectious anemia virus by hypericin, *Photochemistry and Photobiology* 53, 169–74; hu-

man immunodeficiency virus: S. Degar et al., 1992, Inactivation of the human immunodeficiency virus by hypericin: evidence for photochemical alterations of p24 and a block in uncoating, *AIDS Research and Human Retroviruses* 8, 1929–36.

3. R. M. Gulick et al., 1999, Phase I studies of hypericin, the active compound in St. John's Wort, as an antiretroviral agent in HIV-infected adults. AIDS Clinical Trials Group Protocols 150 and 258, *Annals of Internal Medicine* 130, 510–14; J. Lenard et al., 1993, Photodynamic inactivation of infectivity of human immunodeficiency virus and other enveloped viruses using hypericin and rose bengal: Inhibition of fusion and syncytia formation, *Proceedings of the National Academy of Sciences, USA* 90, 158–62; W. Cooper and J. James, 1990, An observational study of the safety and efficacy of hypericin in HIV+ subjects, *International Conference on AIDS* 6, 369; G. Lavie et al., 1995, Hypericin as an inactivator of infectious viruses in blood components, *Transfusion* 35, 392–400; A. Steinbeck-Klose and P. Wernet, 1993, Successful long term treatment over forty months of HIV-patients with intravenous Hypericin, *International Conference on AIDS* 9 (1), 470. Recently, Piscitelli et al. showed that St John's wort interferes with the action of indinavir, an HIV-1 protease inhibitor. R. C. Piscitelli et al., 2000, Indinavir concentrations and St John's wort, *The Lancet* 355, 547–48.

4. A. L. Miller, 1998, St. John's Wort *(Hypericum perforatum)*: Clinical effects on depression and other conditions, *Alternative Medicine Review* 3, 18–26; K. Linde et al., 1996, St John's wort for depression: An overview and meta-analysis of randomised clinical trials, *British Medical Journal* 313, 253–58; M. Philipp et al., 1999, *Hypericum* extract versus imipramine or placebo in patients with moderate depression: Randomised multicentre study of treatment for eight weeks, *British Medical Journal* 319, 1534–38.

5. D. A. Bennet et al., 1998, Neuropharmacology of St. John's Wort *(Hypericum), Annals of Pharmacotherapy* 32, 1201–8.

6. J. M. Cott, 1997, In vitro receptor binding and enzyme inhibition by *Hypericum perforatum* extract, *Pharmacopsychiatry* 30 (suppl. 2), 108–12; K. H. Baureithel et al., 1997, Inhibition of benzodiazepine binding in vitro by amentoflavone, a constituent of various species of *Hypericum, Pharmaceutica Acta Helvetiae* 72, 153–57.

7. S. H. Barondes, 1993, *Molecules and Mental Illness*, New York, Freeman.

8. P. A. DeSmet and W. A. Noten, 1996, St. John's wort as an antidepressant, *British Medical Journal* 313, 241–42. For example, a recent study (G. Harrer et al., 1999, Comparison of equivalence between the St. John's wort extract LoHyp-57 and fluoxetine, *Arzneimittelforschung* 49, 289–96) showed that a St. John's wort extract (LoHyp-57, 800 mg per day) was as effective as low-dose Prozac (20 mg per day) in elderly patients with mild or moderate depression.

9. C. Thomas and R. S. Pardini, 1992, Oxygen dependence of hypericin-induced phototoxicity to EMT6 mouse mammary carcinoma cells, *Photochemistry and Photobiology* 55, 831–37; Z. Diwu, 1995, Novel therapeutic and diagnostic applications of hypocrellins and hypericins, *Photochemistry and Photobiology* 61, 529–39.

10. L. Margulis et al., 1998, *Five Kingdoms: An Illustrated Guide to the Phyla of Life on Earth*, New York, Freeman.

11. H. S. Jennings, 1906, *Behavior of the Lower Organisms*, New York, Columbia University Press; S. O. Mast, 1911, *Light and the Behavior of Organisms*, New York, Wiley.

12. H.-W. Kuhlmann, 1998, Photomovements in ciliated protozoa, *Naturwissenschaften* 85, 143–54.

13. V. Tartar, 1961, *Biology of Stentor*, New York, Pergamon.

14. R. Lattimore (translator), 1971, *The Iliad of Homer*, Chicago, University of Chicago Press.

15. Huang and Pitelka, 1973, The contractile process in the ciliate, *Stentor coeruleus*, part 1, The role of microtubules and filaments, *Journal of Cell Biology* 57, 704–28.

16. A. C. Giese, 1973, *Blepharisma: The Biology of a Light-Sensitive Protozoan*, Stanford, Stanford University Press.

17. G. Checcucci et al., 1997, Chemical structure of blepharismin, the photosensor pigment of *Blepharisma japonicum*, *Journal of the American Chemical Society* 119, 5762–63.

18. Jennings, 1906; Mast, 1911.

19. A related species, *Stentor niger*, exhibits positive phototaxis, movement toward a light source. M. Tuffrau, 1957, Les facteurs essentiels du phototropisme chez le cilié hétérotriche *Stentor niger*, *Bulletin Societe Zoologie de France* 82, 354–56.

20. P. S. Song et al., 1991, Photoreception and photomovement in *Stentor coeruleus*, pp. 267–79 in *Biophysics of Photoreceptors and Photomovements in Microorganisms* (F. Lenci et al., eds.), New York, Plenum.

21. H. Machemer and P. F. M. Teunis, 1996, Sensory-motor coupling and motor responses, pp. 379–402 in *Ciliates: Cells as Organisms* (K. Hausmann and P. C. Bradbury, eds.), Amsterdam, Lubrecht and Cramer.

22. F. Ghetti et al., 1992, Photosensitized reactions as primary molecular events in photomovements of microorganisms, *Journal of Photochemistry and Photobiology B* 15, 185–98.

23. Gulick et al., 1999.

24. Giese, 1973.

25. A. C. Giese, 1981, The photobiology of Blepharisma, pp. 139–80 in: *Photochemical and Photobiological Reviews* (K. C. Smith, ed.), New York, Plenum.

26. P. S. Song et al., 1983, The photoreceptor in *Stentor coeruleus*, pp. 503–20 in *The Biology of Photoreception* (D. J. Cosens and D. Vince-Prue, eds.), New York, Cambridge University Press.

27. R. Dai et al., 1995, Initial spectroscopic characterization of the ciliate photoreceptor stentorin, *Biochimica Biophysica Acta* 1231, 58–68.

28. P. S. Song, 1995, The photo-mechanical responses in the unicellular ciliates, *Journal of Photoscience* 2, 31–35.

29. D. Gioffre et al., 1993, Isolation and characterization of the presumed photoreceptor protein of *Blepharisma japonicum*, *Photochemistry and Photobiology* 58, 275–79; G. Checcucci et al., 1997, Chemical structure of Blepharismin, the photosensor pigment for *Blepharisma japonicum*, *Journal of the American Chemical Society* 119, 5762–63. For peaks, see P. Scevoli et al., 1987, Photomotile responses of *Blepharisma japonicum*, *Journal of Photochemistry and Photobiology* B 1, 75–84.

30. Song, 1995.

31. D. Wood, 1976, Action spectrum and electrophysiological responses correlated with the photophobic response of *Stentor coeruleus*, *Photochemistry and Photobiology* 24, 261–66; T. A. Wells et al., 1997, Electron transfer quenching and the photoinduced EPR of hypericin and the ciliate photoreceptor stentorin, *Journal of Physical Chemistry* 101, 366–72; H. Fabczak et al., 1993a, Photosensory transduction in ciliates: Role of intracellular pH and comparison between *Stentor coeruleus* and *Blephar-*

isma japonicum, Journal of Photochemistry and Photobiology B 21, 47–52; T. Matsuoka et al., 1992, Photoreceptor pigment in *Blepharisma:* H+ release from red pigment, *Photochemistry and Photobiology* 56, 399–402.

32. H. Fabczak et al., 1993a; S. Fabczak et al., 1993a, Photosensory transduction in ciliates, part 1, An analysis of light-induced electrical and motile responses in *Stentor coeruleus, Photochemistry and Photobiology* 57, 696–701; S. Fabczak et al., 1993c, Photosensory transduction in ciliates, part 3, The temporal relation between membrane potentials and photomotile responses in *Blepharisma japonicum, Photochemistry and Photobiology* 57, 872–76.

33. Dai et al., 1995.

34. S. Fabczak et al., 1993a; S. Fabczak et al., 1993c.

35. Wood, 1976.

36. A. Fein and E. Z. Szuts, 1982, *Photoreceptors: Their Role in Vision,* New York, Cambridge University Press. In contrast, vertebrate rod and cone cells hyperpolarize (become more negative) upon illumination.

37. Song, 1995. An alternative hypothesis suggests that electron transfer from the hypericin-like pigments plays an important role. Wells et al., 1997.

38. S. Fabczak et al., 1993b, Photosensory transduction in ciliates, part 2, Possible role of G-protein and cGMP in *Stentor coeruleus, Photochemistry and Photobiology* 57, 702–6; H. Fabczak et al., 1993b, Photosensory transduction in ciliates, part 4, Modulation of the photomovement response of *Blepharisma japonicum* by cGMP, *Photochemistry and Photobiology* 57, 889–92.

39. B. Rayer et al., 1990, Phototransduction: Different mechanisms in vertebrates and invertebrates, *Journal of Photochemistry and Photobiology* B 7, 107–48.

Notes to Sidebars

° The plant acquired its name because it blooms around June 24, the date of the feast of Saint John; wort comes from the Old English *wyrt,* herb.

† United States Department of Agriculture, 1970, *Selected Weeds of the United States,* Washington, D.C., Government Printing Office.

‡ A. C. Giese, 1980, Hypericism, pp. 229–55 in *Photochemical and Photobio-*

logical Review, vol. 6 (K. C. Smith, ed.), New York, Plenum; N. Duran and P. Song, 1986, Hypericin and its photodynamic action, *Photochemistry and Photobiology* 43, 677–80.

§ J. H. Millspaugh, 1897, *Medicinal Plants,* New York, Dover rpt., 1974.

J. Gerard, 1597, *The Herball or generall historie of plantes,* New York, Dover rpt., 1964.

°° B. Diehn et al., 1977, Terminology of behavioral responses of motile microorganisms, *Photochemistry and Photobiology* 26, 559–60.

12. Turning on a Butterfly

1. L. v. Salvini-Plawen and E. Mayr, 1977, On the evolution of photoreceptors and eyes, *Evolutionary Biology* 10, 207–63; J. J. Wolken, 1988, Photobehavior of marine invertebrates: Extraocular photoreception, *Comparative Biochemistry and Physiology* C 91, 145–49.

2. D. Nichols, 1975, *The Uniqueness of Echinoderms,* New York, Oxford University Press.

3. H. Meissl, 1997, Photic regulation of pineal function: Analogies between retinal and pineal photoreception, *Biology of the Cell* 89, 549–54; S. Yokoyama, 1996, Molecular evolution of retinal and nonretinal opsins, *Genes to Cells* 1, 787–94.

4. P. Rohlich et al., 1970, Fine structure of photoreceptor cells in the earthworm, Lumbricus terrestris, *Zeitschriftfur Zellforschung und Mikroskopische Anatomie* 104, 345–57; C. A. Edwards and J. R. Lofty, 1972, *Biology of Earthworms,* New York, Halsted.

5. W. G. Eberhard, 1985, *Sexual Selection and Animal Genitalia,* Cambridge, Harvard University Press; W. G. Eberhard, 1990, Animal genitalia and female choice, *American Scientist* 78, 134–41.

6. Ibid.; R. A. Fortey and R. H. Thomas, 1998, *Arthropod Relationships,* London, Chapman and Hall.

7. Eberhard, 1985; Eberhard, 1990.

8. Ibid.

9. Eberhard, 1985.

10. Ibid.; Eberhard, 1990.

11. W. G. Eberhard, 1996, *Female Control: Sexual Selection by Cryptic Female Choice,* Princeton, Princeton University Press.

12. C. Darwin, 1871, *Descent of Man, and Selection in Relation to Sex,* Princeton, Princeton University Press (1981 rpt.).

13. Eberhard, 1985; Eberhard, 1990.

14. M. Andersson, 1994, *Sexual Selection,* Princeton, Princeton University Press; J. L. Gould and C. G. Gould, 1989, *Sexual Selection: Mate Choice and Courtship in Nature,* New York, Freeman. "Direct selection" is an alternative mechanism that could lead to elaborate male displays and extreme female preferences. The relative significance of direct and indirect selection in generating female preference and male displays is controversial. See M. Kirkpatrick and M. J. Ryan, 1991, The evolution of mating preferences and the paradox of the lek, *Nature* 350, 33–38; M. Kirkpatrick, 1996, Good genes and direct selection in the evolution of mating preferences, *Evolution* 50, 2125–40; Andersson, 1994; A. Zahavi and A. Zahavi, 1997, *The Handicap Principle: A Missing Piece of Darwin's Puzzle,* New York, Oxford University Press.

15. Eberhard, 1985; Eberhard, 1990.

16. Ibid.

17. D. Suyama et al., 1994, The genital photoreceptor of the male butterfly is necessary for establishment of the copulation, *Zoological Society of Japan,* abstract.

18. Andersson, 1994.

19. A. F. Dixson, 1987, Observations on the evolution of genitalia and copulatory behavior in primates, *Journal of Zoology* 213, 423–43.

20. K. Arikawa et al., 1980, Multiple extraocular photoreceptive areas on genitalia of butterfly *Papilio xuthus, Nature* 288, 700–702.

21. K Arikawa and Y. Miyako-Shimazaki, 1996, Combination of physiological and anatomical methods for studying extraocular photoreceptors on the genitalia of the butterfly, *Papilio xuthus, Journal of Neuroscience Methods* 69, 75–82.

22. T. Okano et al., 1995, Molecular basis for tetrachromatic color vision, *Comparative Biochemistry and Physiology B Biochemistry and Molecular Biology* 112, 405–14; A. T. Bennett and I. C. Cuthill, 1994, Ultraviolet vision in birds: What is its function? *Vision Research* 34, 1471–78.

23. K. Arikawa et al., 1987, Pentachromatic visual system in a butterfly, *Naturwissenschaften* 74, 297–98; K. Bandai et al., 1992, Localization of spectral receptors in the ommatidium of butterfly compound eye deter-

mined by polarization sensitivity, *Journal of Comparative Physiology* A 171, 289–97; M. Kinoshita, 1997, Spectral receptors of nymphalid butterflies, *Naturwissenschaften* 84, 199–201; K. Arikawa, 1997, Color vision, in *Atlas of Arthropod Sensory Receptors* (E. Eguchi and Y. Tominaga, eds.), Tokyo, Springer-Verlag. A. D. Briscoe (1998, Molecular diversity of visual pigments in the butterfly Papilio glaucus, *Naturwissenschaften* 85, 33–35) has cloned fragments of six opsin genes using mRNA isolated from the head of a single female of a related species, *Papilio glaucus*.

24. M. Kinoshita et al., 1999, Colour vision of the foraging swallowtail butterfly *Papilio xuthus*, *Journal of Experimental Biology* 202, 95.
25. Arikawa et al., 1987; K. Bandai et al., 1992.
26. Arikawa et al., 1987.
27. Arikawa et al., 1980; Arikawa and K. Aoki et al., 1984, Inhibitory effect of genital photoreceptors on the activity of abdominal motoneurones in *Papilio xuthus* (Lepidoptera, Papilionidae), *Journal of Insect Physiology* 30, 853–60.
28. Arikawa et al., 1980; K. Arikawa et al., 1991, Extraocular photoreceptors in the last abdominal ganglion of a butterfly, *Papilio xuthus*, *Naturwissenschaften* 78, 82–84.
29. Arikawa et al., 1980; Y. Miyako et al., 1993, Ultrastructure of the extraocular photoreceptor in the genitalia of a butterfly, *Papilio xuthus*, *Journal of Comparative Neurology* 327, 458–68.
30. Ibid.
31. Arikawa et al., 1980.
32. Salvini-Plawen and Mayr, 1977.
33. Arikawa and Aoki, 1984.
34. Suyama et al., 1994.
35. Arikawa and Aoki, 1984; K Arikawa et al., 1996, Light on butterfly mating, *Nature* 382, 119.
36. K. Arikawa, 1993, Valva-opening response induced by the light stimulation of the genital photoreceptors of male butterflies, *Naturwissenschaften* 80, 326–28.
37. Arikawa et al., 1996; K. Arikawa et al., 1997, Hindsight by genitalia: Photo-guided copulation in butterflies, *Journal of Comparative Physiology* A 180, 295–99.
38. Arikawa et al., 1980.

13. Blue Moons and Red Tides

1. W. M. Porch, 1989, Blue moons and large fires, *Applied Optics* 28, 1778–84.

2. M. Henry, 1998, Red Tide Chronology (www.marinelab.sarasota.fl.us/
~mhenry/rtchrono.phtml).

3. T. Simkin and R. Fiske, 1983, *Krakatau, 1883,* Washington, D.C., Smithsonian Institution Press; I. Thornton, 1996, *Krakatau: The Destruction and
Reassembly of an Island Ecosystem,* Cambridge, Harvard University Press.

4. Porch, 1989; M. J. G. Minnaert, 1974, *Light and Color in the Outdoors,*
New York, Springer-Verlag (trans. and rev. by Len Seymour, 1993).

5. Porch, 1989.

6. Anonymous, 1998, Blue Sky (The Exporatorium) (www.explorato-
rium.edu/snacks/blue_sky.html); Minnaert, 1993.

7. Minnaert, 1974; Porch, 1989.

8. Apparently, this phrase, meaning "very rarely," is not etymologically related to the rarity of the blue moon optical effect caused by volcanic dust.
The *OED* cites a usage of the phrase "blue moon" in 1528: "Yf they saye
the mone is belewe, We must beleve that it is true." In spite of this early
usage, Pierce Egan included an explanatory footnote for his sentence
"Haven't seen you this blue moon" in *Real Life in London,* published in
1821 (cited in *OED;* personal communication with Margot Carlton, *Oxford English Dictionary,* June 22, 1998). A more modern usage of the
phrase "blue moon" to refer to the second full moon of a month (a rare
event) presumably derives from a misinterpretation of the *Maine
Farmer's Almanac* by J. Hugh Pruett in an article he wrote for *Sky and
Telescope* in 1937 (D. W. Olson et al., 1999, What's a blue moon? *Sky and
Telescope* 97 [5], 23–27).

9. B. W. Halstead, 1991, *Dangerous Aquatic Animals of the World,* Princeton,
Darwin.

10. D. M. Anderson, 1997, Turning back the harmful red tide, *Nature* 388,
513–14; D. M. Anderson, 1994, Red tides, *Scientific American* 271 (2),
52–58.

11. M. Moore-Ede, et al., 1982, *The Clocks That Time Us,* Cambridge, Harvard University Press.

12. M. De Mairan (ed.), 1729, Observation botanique, *Historie de l'Academie
Royale des Sciences,* Paris.

13. F. B. Salisbury and C. W. Ross, 1991, *Plant Physiology*, Belmont, Calif., Wadsworth.

14. Alexis P. de Candole first demonstrated in 1832 that Mimosa leaf movements are not exactly twenty-four hours when plants are held under constant conditions. S. S. Golden, 1997, Cyanobacterial circadian rhythms, *Annual Review of Plant Physiology and Plant Molecular Biology* 48, 327–54.

15. Unicellular organisms: R. Pohl, 1948, Tagesrhythmus in phototaktischem Verhalten der *Euglena gracilis*, *Zeitschrift Naturforschung* 3, 367–74; bacteria: B. M. Sweeney and M. B. Borgese, 1989, A circadian rhythm in cell division in a prokaryote, the cyanobacterium *Synechoccus* WH7803, *Journal of Phycology* 25, 183–86; B. Bergman et al., 1997, Nitrogen fixing non-heterocystous cyanobacteria, *FEMS Microbiology Reviews* 19, 139–85; C. H. Johnson et al., 1996, Circadian clocks in prokaryotes, *Molecular Microbiology* 21, 5–11; D. Lloyd, 1998, Circadian and ultradian clock-controlled rhythms in unicellular microorganisms, *Advances in Microbial Physiology* 39, 291–38.

16. Rhodopsin: Moore-Ede et al., 1982; flavoprotein: Y. Miyamoto and A. Sancar, 1998, Vitamin B2-based blue-light photoreceptors in the retinohypothalamic tract as the photoactive pigments for setting the circadian clock in mammals, *Proceedings of the National Academy of Sciences, USA* 95, 6097–102; R. J. Thresher et al., 1998, Role of mouse cryptochrome blue-light photoreceptor in circadian photoresponses, *Science* 282, 1490–94.

17. G. Fritz et al., 1989, Free cellular riboflavin is involved in phase shifting by light of the circadian clock in *Neurospora crassa*, *Plant and Cell Physiology* 30, 557–64.

18. Phytochrome: R. Sater and A. W. Galston, 1981, Mechanisms of controls of leaf movements, *Annual Review of Plant Physiology* 32, 82–110; flavoprotein: A. J. Millar et al., 1995, The regulation of circadian period by phototransduction pathways in Arabidopsis, *Science* 267, 1163–66.

19. M. Mittag and J. W. Hastings, 1996, Exploring the signalling pathway of circadian bioluminescence, *Physiologia Plantarum* 96, 727–32; C. S. Pittendrigh, 1993, Temporal organization: Reflections of a Darwinian clockwatcher, *Annual Review of Physiology* 55, 17–54.

20. L. Wetterberg, 1993, *Light and Biological Rhythms in Man*, New York,

Pergamon; M. S. Shafii and S. L. Shafii, 1990, *Biological Rhythms, Mood Disorders, Light Therapy and the Pineal Gland,* Washington, D.C., American Psychiatric Press.

21. T. Roenneberg, 1996, The complex circadian system of *Gonyaulax polyedra, Physiologia Plantarum* 96, 733–37; Mittag and Hastings, 1996.

22. Y. Inagaki et al., 1997, Algae or protozoa: phylogenetic position of euglenophytes and dinoflagellates as inferred from mitochondrial sequences, *Journal of Molecular Evolution* 45, 295–300.

23. R. Krasnow et al., 1980, Circadian spontaneous bioluminescent glow and flashing of *Gonyaulax polyedra, Journal of Comparative Physiology* 138, 19–26.

24. M. Desjardins and D. Morse, 1993, The polypeptide components of scintillons, the bioluminescence organelles of the dinoflagellate Gonyaulax polyedra, *Biochemistry and Cellular Biology* 71, 176–82.

25. L. Fritz et al., 1990, The circadian bioluminescence rhythm of Gonyaulax is related to daily variations in the number of light-emitting organelles, *Journal of Cell Science* 95, 321–28. Expression of proteins involved in bioluminescence is regulated at the translational level. Mittag and Hastings, 1996; S. Morse et al., 1990, What is the clock? Translational regulation of circadian bioluminescence, *Trends in Biochemical Science* 15, 262–65; P. Milos et al., 1990, Circadian control over synthesis of many Gonyaulax proteins is at a translational level, *Naturwissenschaften* 77, 87–89.

26. Mittag and Hastings, 1996; Roenneberg, 1996.

27. T. Roenneberg and J. W. Hastings, 1988, Two photoreceptors control the circadian clock of a unicellular alga, *Naturwissenschaften* 75, 206–7; D. Morse et al., 1994, Different phase responses of the two circadian oscillators in *Gonyaulax, Journal of Biological Rhythms* 9, 263–74.

28. C. H. Johnson and J. W. Hastings, 1989, Circadian phototransduction: phase resetting and frequency of the circadian clock of *Gonyaulax* cells in red light, *Journal of Biological Rhythms* 4, 417–37. The action spectrum for phase-shifting of the bioluminescence rhythm has maxima near 475 nm (blue light) and 650 nm (red light), and a minimum between 500–550 nm (green light), similar to the action spectrum for photosynthesis. J. W. Hastings and B. Sweeney, 1960, The action spectrum for shifting the phase of the rhythm of luminescence in *Gonyaulax polyedra, Journal of General Physiology* 43, 697–706.

29. Roenneberg and Hastings, 1988.
30. B. Poeggeler et al., 1991, Pineal hormone melatonin oscillates also in the dinoflagellate *Gonyaulax polyedra*, *Naturwissenschaften* 78, 268–69; R. Hardeland et al., 1995, On the primary functions of melatonin in evolution: Mediation of photoperiodic signals in a unicell, photooxidation, and scavenging of free radicals, *Journal of Pineal Research* 18, 104.
31. I. Balzer and R. Hardeland, 1996, Melatonin in algae and higher plants: Possible new roles as a phytohormone and antioxidant, *Botanica Acta* 109, 180–83.
32. L. Glass and M. C. Mackey, 1988, *From Clocks to Chaos*, Princeton, Princeton University Press; J. E. Muller et al., 1985, Circadian variation in the frequency of onset of acute myocardial infarction, *New England Journal of Medicine* 313, 1315–22.

Notes to Sidebars

° J. W. Hastings, 1995, Bioluminescence: Similar chemistries but many different evolutionary origins, *Photochemistry and Photobiology* 62, 599–600.

† J. W. Hastings, 1983, Biological diversity, chemical mechanisms, and the evolutionary origins of bioluminescent systems, *Journal of Molecular Evolution* 19, 309–21; Hastings, 1995; J. W. Hastings, 1996, Chemistries and colors of bioluminescent reactions: a review, *Gene* 173, 5–11.

‡ J. C. Dunlap, 1996, Genetic and molecular analysis of circadian rhythms, *Annual Review of Genetics* 30, 579–601.

§ A. Goldbetter, 1996, *Biochemical Oscillations and Cellular Rhythms*, New York, Cambridge University Press; T. Pavlidis, 1973, *Biological Oscillators: Their Mathematical Analysis*, New York, Academic Press; A. T. Winfree, 1980, *The Geometry of Biological Time*, New York, Springer-Verlag; A. T. Winfree, 1987, *Timing of Biological Clocks*, New York, Freeman; J. C. Dunlap, 1999, Molecular bases for circadian clocks, *Cell* 96, 271–90.

T. Pavlidis, 1973.

°° Dunlap, 1999; S. K. Crosthwaite et al., 1997, Neurospora wc-1 and wc-2: transcription, photoresponses, and the origins of circadian rhythmicity, *Science* 276, 763–69; D. P. King et al., 1997, Positional cloning of the mouse circadian clock gene, *Cell,* 89, 641–53; H. Tei et al., 1997, Circadian oscillation of a mammalian homologue of the Drosophila period gene, *Nature* 389, 512–16.

14. Photosynthesis and the Great Salt Lake

1. E. G. Ruestow, 1996, *The Microscope in the Dutch Republic: The Shaping of Discovery,* New York, Cambridge University Press; P. De Kruif, 1926, *Microbe Hunters,* New York, Harcourt, Brace (1996 rpt.).

2. L. Margulis et al., 1998, *Five Kingdoms: An Illustrated Guide to the Phyla of Life on Earth,* New York, Freeman.

3. E. I. Friedmann et al., 1993, Long-term productivity in the cryptoendolithic microbial community of the Ross Desert, Antarctica, *Microbial Ecology* 25, 51–69; E. I. Friedmann (ed.), 1993, *Antarctic Microbiology,* New York, Wiley-Liss.

4. C. R. Woese and G. E. Fox, 1977, Phylogenetic structure of the prokaryotic domain: The primary kingdoms, *Proceedings of the National Academy of Science, USA* 74, 5088–90.

5. While Woese and colleagues are credited with reclassification of prokaryotes now called Archaea, halophilic (salt-loving) species in this group have long been known to cause the reddening and spoilage of salted meats. G. J. Banwart, 1989, *Basic Food Microbiology,* New York, Van Nostrand Reinhold. This was probably known even by prehistoric humans. It has also probably long been known that halophilic organisms cause natural salterns to turn red over time.

6. S. M. Barnes et al., 1994, Remarkable archael diversity detected in a Yellowstone National Park hot spring environment, *Proceedings of the National Academy of Science, USA* 91, 1609–13; E. F. DeLong, 1992, Archaea in coastal marine environments, *Proceedings of the National Academy of Science, USA* 89, 5685–89; J. L. Stein, 1996, Characterization of uncultivated prokaryotes: Isolation and analysis of a 40-kilobase-pair genome fragment from a planktonic marine archaeon, *Journal of Bacteriology* 178, 591–99.

7. Lynn Margulis, who has proposed a symbiosis-based classification system (L. Margulis, 1996, Archaeal-eubacterial mergers in the origin of

Eukarya: Phylogenetic classification of life, *Proceedings of the National Academy of Science, USA* 93, 1071–76), is a notable exception.

8. N. Iwabe et al., 1989, Evolutionary relationship of archaebacteria, eubacteria, and eukaryotes inferred from phylogenetic trees of duplicated genes, *Proceedings of the National Academy of Science, USA* 86, 9355–59; J. P. Gogarten et al., 1989, Evolution of the vacuolar H+-ATPase: implications for the origin of eukaryotes, *Proceedings of the National Academy of Science, USA* 86, 6661–65; C. Woese et al., 1990, Towards a natural system of organisms: Proposal for the domains Archaea, Bacteria, and Eucarya, *Proceedings of the National Academy of Science, USA* 87, 4576–79; J. R. Brown and W. F. Doolittle, 1995, Root of the universal tree of life based on ancient aminoacyl-tRNA synthetase gene duplications, *Proceedings of the National Academy of Science, USA* 92, 2441–45; S. Baldauf et al., 1996, The root of the universal tree and the origin of eukaryotes based on elongation factor phylogeny, *Proceedings of the National Academy of Science, USA* 93, 7749–54.

9. This species was previously called *Halobacterium halobium.*

10. D. Petracchi et al., 1994, Photobehaviour of *Halobacterium halobium:* proposed models for signal transduction and motor switching, *Journal of Photochemistry and Photobiology* B 24, 75–99.

11. M. Kates et al. (eds.), 1993, *The Biochemistry of Archaea (Archaebacteria),* New York, Elsevier; D.-P. Häder, 1987, Photosensory behavior in procaryotes, *Microbiology Review* 51, 1–21.

12. R. R. Birge, 1994, A nonlinear proton pump, *Nature* 371, 659–60.

13. W. Stoeckenius et al., 1967, A morphological study of Halobacterium halobium and its lysis in media of low salt concentration, *Journal of Cell Biology* 34, 365–93.

14. M. P. Krebs and G. Khorana, 1993, Mechanism of light-dependent proton translocation by bacteriorhodopsin, *Journal of Bacteriology* 175, 1555–60. Strictly speaking, vitamin A is retinol or retinoids that have the same biological activities as retinol. R. Blumhoff et al., 1992, Vitamin A: Physiological and biochemical processing, *Annual Review of Nutrition* 12, 37–57; R. Blumhoff et al., 1991, Vitamin A: new perspectives on absorption, transport, and storage, *Physiological Reviews* 71, 951–90; *Physiological Reviews* 71, 951–90; R. Blomhoff et al., 1990, Transport and storage of vitamin A, *Science* 250, 399–404.

15. D. Oesterhelt and W. Stoeckenius, 1973, Functions of a new photoreceptor membrane, *Proceedings of the National Academy of Science, USA* 70, 2853–57.

16. D. Osterhelt and W. Stoeckenius, 1971, Rhodopsin-like protein from the purple membrane of *Halobacterium halobium, Nature, New Biology* 233, 149–52.

17. M. S. Braiman et al., 1988, Vibrational spectroscopy of bacteriorhodopsin mutants: Light-driven proton transport involves protonation changes of aspartic acid residues 85, 96, and 212, *Biochemistry* 27, 8516–20; K. Gerwert et al., 1989, Role of aspartate-96 in proton translocation by bacteriorhodopsin, *Proceedings of the National Academy of Science, USA* 86, 4943–47.

18. P. Mitchell, 1979, Keilin's respiratory chain concept and its chemiosmotic consequences, *Science* 206, 1148–59.

19. F. Gai et al., 1998, Chemical dynamics in proteins: The photoisomerization of retinal in bacteriorhodopsin, *Science* 279, 1886–91.

20. Common techniques for probing the electronic state of proteins include time-resolved changes in absorption, fluorescence, resonance Raman spectroscopy, and the recently developed technique of time-resolved X-ray crystallography. Gai et al., 1998; B. Perman et al., 1998, Energy transduction on the nanosecond time scale: early structural events in a xanthopsin photocycle, *Science* 279, 1946–50.

21. J. K Lanyi, 1997, Mechanism of ion transport across membranes: Bacteriorhodopsin as a prototype for proton pumps, *Journal of Biological Chemistry* 272, 31209–12.

22. A. Matsuno-Yagi and Y. Mukohata, 1977, Two possible roles of bacteriorhodopsin: A comparative study of strains of *Halobacterium halobium* differing in pigmentation, *Biochemical and Biophysical Research Communications* 78, 237–43.

23. B. Schobert and J. K. Lanyi, 1982, Halorhodopsin is a light-driven chloride pump, *Journal of Biological Chemistry* 257, 10306–13. See also E. V. Lindley and R. E. MacDonald, 1979, A second mechanism for sodium extrusion in *Halobacterium halobium:* A light-driven sodium pump, *Biochemical and Biophysical Research Communications* 88, 491–99; R. E. MacDonald et al., 1979, Characterization of the light-driven sodium pump of *Halobacterium halobium:* Consequences of sodium efflux

as the primary light-driven event, *Journal of Biological Chemistry* 254, 11831–38.

24. J. K. Lanyi, 1990, Halorhodopsin, a light-driven electrogenic chloride-transport system, *Physiological Reviews* 70, 319–30.

25. R. A. Bogomolni and J. N. Spudich, 1982, Identification of a third rhodopsin-like pigment in phototactic *Halobacterium halobium*, *Proceedings of the National Academy of Sciences, USA* 79, 6250–54; E. N. Spudich and J. L. Spudich, 1982, Control of transmembrane ion fluxes to select halorhodopsin-deficient and other energy-transduction mutants of *Halobacterium halobium*, *Proceedings of the National Academy of Sciences, USA* 79, 4308–12.

26. J. L. Spudich and R. A. Bogomolni, 1988, Sensory rhodopsins of halobacteria, *Annual Review of Biophysics and Biophysical Chemistry* 17, 193–215. Like halorhodopsin and bacteriorhodopsin, the sensory rhodopsins use all-trans retinal as a chromophore, and this undergoes isomerization to 13-cis retinal. The different amino acid sequences of these different proteins determines their different absorption spectra. K. Nakanishi, 1991, Why 11-cis retinal? *American Zoologist* 31, 479–89.

27. W. D. Hoff, et al., 1997, Molecular mechanism of photosignaling by ar-chaeal sensory rhodopsins, *Annual Review of Biophysics and Biomolecular Structure* 26, 223–58.

28. H. Berg, 1993, *Random Walks in Biology,* Princeton, Princeton University Press.

29. S. I. Bibikov et al., 1993, Bacteriorhodopsin is involved in halobacterial photoreception, *Proceedings of the National Academy of Sciences, USA* 90, 9446–50; S. I. Bibikov et al., 1991, The proton pump bacteriorhodopsin is a photoreceptor for signal transduction in *Halobacterium halobium, Federation of the European Biochemical Society Letters* 295, 223–26; B. Yan et al., 1992, Transformation of a bop-hop-sop-I-sop-II-*Halobacterium halobium* mutant to bop+: Effects of bacteriorhodopsin photoactivation on cellular proton fluxes and swimming behavior, *Photochemistry and Photobiology* 56, 553–61.

30. K. Ihara et al., 1999, Evolution of the archaeal rhodopsins: Evolution rate changes by gene duplication and functional differentiation, *Journal of Molecular Biology* 285, 163–74.

31. Nakanishi, 1991.

32. It should be noted that some aquatic species use 3,4-dehydroretinal, and insects use 3- or 4-hydroxylretinal. Ibid.

33. M. Delbrück, 1976, Light and life, *Carlsberg Laboratory Research Communications* 41, 299–309; E. D. Lipson and B. A. Horwitz, 1991, Photosensory Reception and Transduction, pp. 1–64 in: *Sensory receptors and Signal Transduction* (J. L. Spudich and B. H. Satir, eds.), New York, Wiley.

34. R. Blomhoff et al., 1991, Vitamin A metabolism: New perspectives on absorption, transport, and storage, *Physiological Reviews* 71, 951–90.

Notes to Sidebar

* W. M. Moreau, 1988, *Semiconductor Lithography: Principles, Practices, and Materials, (Microdevices: Physics and Fabrication Technologies)*, New York, Plenum; J. R. Sheats and B. W. Smith (eds.), 1998, *Microlithography: Science and Technology*, New York, Marcel Dekker.

† D. Oesterhelt et al., 1991, Bacteriorhodopsin: A biological material for information processing, *Quarterly Review of Biophysics* 24, 425–78; R. R. Birge, 1992, Protein-based optical computing and optical memories, *Computer* 25, 56–67.

‡ D. A. Parthenopoulos and P. M. Rentzepis, 1989, Three-dimensional optical storage memory, *Science* 245, 843–45.

§ R. R. Birge, 1994, Protein-based three-dimensional memory, *American Scientist* 82, 348–55; R. R. Birge et al., 1997, Protein-based three-dimensional memories and associative processors, pp. 439–71 in *Molecular Electronics: A "Chemistry for the 21st Century" Monograph* (J. Jortner and M. Ratner, eds.), London, Blackwell Science; B. W. Vought and R. R. Birge, 1999, Molecular electronics and hybrid computers, pp. 477–90 in *Wiley Encyclopedia of Electrical and Electronics Engineering* (J. G. Webster, ed.), New York, Wiley.

R. R. Birge, 1995, Protein-based computers, *Scientific American* 83 (3), 90–95.

15. Too Much of a Good Thing

1. R. H. Whittaker estimated world net primary productivity as 170 billion metric tons per year, equivalent to 187 billion English tons per year.

R. H. Whittaker, 1975, *Communities and Ecosystems*, New York, Macmillan.

2. According to the United Nations, the population of the earth reached 6 billion in the year 1999. United Nations Population Fund, 1999 (www.unfpa.org/swp/swpmain.htm). Thus $(187 \times 10^9$ tons of carbohydrate per year)$/(6 \times 10^9$ people) $= 31.2$ tons of carbohydrate per person per year.

3. Whittaker, 1975.

4. F. B. Salisbury and C. W. Ross, 1991, *Plant Physiology*, Belmont, Calif., Wadsworth. This generalization is true at the level of the individual leaf but not at the level of the whole plant, because leaves at the bottom of a plant's canopy are in shade and not in saturation. Moreover, additional stress factors may lower the level at which saturation occurs. A. S. Verhoeven et al., 1999, The xanthophyll cycle and acclimation of *Pinus ponderosa* and *Malva neglecta* to winter stress, *Oecologia* 118, 277–87.

5. What I call toxic forms of oxygen are generally referred to in the scientific literature as reactive oxygen species. This loosely defined term includes different forms of oxygen and compounds of oxygen with hydrogen, chlorine, and nitrogen that have different properties and reactivities. A. U. Khan and T. Wilson, 1995, Reactive oxygen species as cellular messengers, *Current Biology* 2, 437–45.

6. E. Cadenas, 1989, Biochemistry of oxygen toxicity, *Annual Review of Biochemistry* 58, 79–110; G. Moreno et al. (eds.), 1988, *Photosensitization: Molecular, Cellular, and Medical Aspects*, New York, Springer-Verlag; C. S. Foote, A. A. Frimer, 1985, *Singlet Oxygen*, Boca Raton, Fla., CRC Press.

7. E. Cadenas, 1995, Mechanisms of oxygen activation and reactive oxygen species detoxification, pp. 1–61 in *Oxidative Stress and Antioxidant Defenses in Biology* (S. Ahmad, ed.), New York, Chapman and Hall; Cadenas, 1989.

8. K. F. Guy, 1998, Vitamins E plus C are interacting conutrients required for optimal health: A critical and constructive review of epidemiology and supplementation data regarding cardiovascular disease and cancer, *Biofactors* 7, 113–74; S. R. Maxwell and G. Y. Lip, 1997, Free radicals and antioxidants in cardiovascular disease, *British Journal of Clinical Pharmacology* 44, 307–17; N. R. Sahyoun, 1997, Vitamin C: What do we know and how much do we need? *Nutrition* 13, 835–36; G. van Poppel

and H. van den Berg, 1997, Vitamins and cancer, *Cancer Letters* 114, 195–202.

9. Salisbury and Ross, 1991.

10. P. Horton et al., 1996, Regulation of light harvesting in green plants, *Annual Review of Plant Physiology and Plant Molecular Biology* 47, 655–84.

11. A. Young and G. Britton, 1990, Carotenoids and stress, pp. 87–123 in *Stress Responses in Plants: Adaptation and Acclimation Mechanisms* (R. G. Alscher and J. R. Cummings, eds.), New York, Wiley-Liss.

12. Salisbury and Ross, 1991.

13. C. H. Foyer et al., 1994, Photooxidative stress in plants, *Physiologia Plantarum* 92, 696–717.

14. O. Björkman and B. Demmig-Adams, 1994, Regulation of photosynthetic light energy capture, conversion, and dissipation in leaves of higher plants, pp. 18–48 in *Ecophysiology of Photosynthesis* (E.-D. Schulz and M. M. Caldwell, eds.), New York, Springer-Verlag.

15. C. Neubauer and U. Schreiber, 1989, Photochemical and nonphotochemical quenching of chlorophyll fluorescence induced by hydrogen peroxide, *Zeitschrift Naturforschung* 44c, 262–70; C. Neubauer and H. Y. Yamomoto, 1992, Mehler-peroxidase reaction mediates zeaxanthin formation and zeaxanthin-related fluorescence quenching in intact chloroplast, *Plant Physiology* 99, 1354–61.

16. Foyer et al., 1994.

17. M. Havaux and K. K. Niyogi, 1999, The violaxanthin cycle protects plants from photooxidative damage by more than one mechanism, *Proceedings of the National Academy of Sciences, USA* 96, 8762–67. The reaction for this process is: $^1O_2 + Car \rightarrow {}^3O_2 + {}^3Car$; $^3Car \rightarrow Car + heat$, where $^1O_2$ is singlet oxygen; Car is ground-state carotenoid; $^3O_2$ is ground-state (triplet-state) oxygen; and ^3Car is triplet-state carotenoid. R. G. Alscher and J. L. Hess (eds.), 1993, *Antioxidants in Higher Plants*, Boca Raton, Fla., CRC Press. Some researchers have argued that this reaction has no *in vivo* significance for plants. B. Demmig-Adams et al., 1996, Carotenoids 3: *In vivo* function of carotenoids in higher plants, *Federation of the American Society of Experimental Biology Journal* 10, 403–12.

18. B. Demmig-Adams and W. W. Adams, 1996, Mechanism of light-dependent proton translocation by bacteriorhodopsin, *Trends in Plant Science* 1, 21–26.

19. T. G. Owens, 1997, Processing of excitation energy by antenna pigment, pp. 87–112 in *Photosynthesis and the Environment* (N. R. Baker, ed.), New York, Kluwer.

20. Ibid.; H. A. Frank et al., 1994, Photophysics of the carotenoids associated with the xanthophyll cycle in photosynthesis, *Photosynthesis Research* 41, 389–95.

21. W. W. Adams et al., 1995, "Photoinhibition" during winter stress: involvement of sustained xanthophyll cycle-dependent energy dissipation, *Australian Journal of Plant Physiology* 22, 261–76; A. S. Verhoeven et al., 1996, Close relationship between the state of the xanthophyll cycle pigments and photosystem II efficiency during recovery from winter stress, *Physiologia Plantarum* 96, 567–76; Verhoeven et al., 1999.

22. B. Demmig-Adams et al., 1989, Light response of CO_2 assimilation, dissipation of excess excitation energy, and zeaxanthin content of sun and shade leaves, *Plant Physiology* 90, 881–89.

23. Demmig-Adams and Adams, 1996; B. Demmig-Adams and W. W. Adams, 1993, The xanthophyll cycle, pp. 91–110 in *Antioxidants in Higher Plants* (R. G. Alscher and J. L. Hess, eds.), Boca Raton, Fla., CRC Press.

24. W. W. Adams et al., 1992, Leaf orientation and response of the xanthophyll cycle to incident light, *Oecologia* 90, 404–10.

25. Demmig-Adams and Adams, 1996; Demmig-Adams and Adams, 1993.

26. P. Böger and G. Sandmann, 1990, Modern herbicides affecting typical plant processes, pp. 173–216 in *Chemistry of Plant Protection*, vol. 6 (W. S. Bowers et al., eds.), Berlin, Springer-Verlag.

27. B. Demmig-Adams and W. W. Adams, 1992, Carotenoid composition in sun and shade leaves of plants with different life forms, *Plant Cell and Environment* 15, 411–19; S. S. Thayer and O. Björkman, 1990, Leaf xanthophyll content and composition in sun and shade leaves determined by HPLC, *Photosynthesis Research* 23, 331–43.

28. K. K. Niyogi et al., 1998, Arabidopsis mutants define a central role for the xanthophyll cycle in the regulation of photosynthetic energy conversion, *Plant Cell* 10, 1121–34.

29. K. K. Niyogi et al., 1997a, *Chlamydomonas* xanthophyll mutants identified by video imaging of chlorophyll fluorescence quenching, *Plant Cell* 9, 1369–80.

30. K. K. Niyogi et al., 1997b, The roles of specific xanthophylls in photo-protection, *Proceedings of the National Academy of Sciences, USA* 94, 14162–67.

31. Readers may wish to compare my vision to the final paragraph in *The Origin of Species* (C. Darwin, 1859, *The Origin of Species*, London, Murray).

Notes to Sidebars

* K. Asada, 1994, Production and action of active oxygen species in photosynthetic tissues, pp. 77–104 in *Causes of Photooxidative Stress and Amelioration of Defense Systems in Plants* (C. H. Foyer and P. M. Mullineaux, eds.), Boca Raton, Fla., CRC Press.

† T. W. Goodwin, 1980, *Biochemistry of Carotenoids*, New York, Chapman and Hall; G. Britton et al. (eds.), 1995, *Carotenoid*, Berlin, Birkhauser-Verlag.

‡ T. W. Goodwin, 1988, *Plant Pigments*, San Diego, Academic Press; G. Britton, 1983, *The Biochemistry of Natural Pigments*, Berlin, Birkhauser Verlag.

§ I am referring to the C_{40} carotenoids. Carotenoids with more than forty carbon atoms are called homo-carotenoids, and those with fewer than forty are called apo-carotenoids. R. Edge and T. G. Truscott, 2000, Carotenoids: Free radical interactions, *Spectrum* 13 (1), 12–20.

S. Pratt, 1999, Dietary prevention of age-related macular degeneration, *Journal of the Optometry Association* 70, 39–47; Edge and Truscott, 2000.

Glossary

Action spectrum: Graph that shows the effectiveness of different wavelengths of light in eliciting a chemical or biological response like vision, photosynthesis, phototropism, or phototaxis.

Adenosine triphosphate (ATP): Universal energy storage compound that cells use to drive energy-requiring reactions (see figure A10, Appendix).

Air glow: Upper-atmospheric phenomenon caused by the delayed emission of light from nitrogen and oxygen atoms that are energized by sunlight during daytime.

Alpha-acid: Compound present in hops that is transformed into iso-alpha acid during the wort boil of beer production (see figure A7, Appendix).

Archaea: Major group of prokaryotes that are genetically and metabolically different from Eubacteria, including the halophiles, thermophiles, and methanogens (see figures 1, p. 4; and 11, p. 163).

Bacteriorhodopsin: Light-sensitive, retinal-based, membrane-bound protein that pumps protons out of *Halobacterium salinarum* cells and drives the chemiosmotic synthesis of ATP (see figures 12, p. 167; 13, p. 170; and A13, Appendix).

Basal cell carcinoma: Malignant growth of basal cells of epidermal tissue that is the most common form of skin cancer but typically does not metastasize.

Bioluminescence: Production of light from biochemical reactions within a living organism.

Black body: Idealized solid body that emits a continuous spectrum of radiation, with the wavelength maximum moving from the infrared to the ultraviolet as its temperature increases.

Blepharismin: Hypericin-like photoreceptive pigment that controls cellular movement in *Blepharisma japonicum.*

Carotene: Red, orange, or yellow carotenoid pigment that is composed entirely of hydrogen and carbon atoms, typically has nine conjugated double bonds, and is the precursor of vitamin A.

Carotenoid: Red, orange, or yellow fat-soluble pigment that is synthesized from eight 5-carbon isoprene molecules and is classified as a carotene or xanthophyll.

Chemiosmosis: Biochemical mechanism in which flow of an intermediate species, such as a proton (H^+), across a membrane is used to drive another reaction, such as the synthesis of ATP (see figures 12, p. 167; 13, p. 170; A13, Appendix).

Chemotaxis: Orientation movement by a motile organism in response to a chemical.

Chlorofluorocarbon (CFC): One of a family of chemicals that is composed of carbon, hydrogen, chlorine, and fluorine atoms and can be used as a refrigerant or electronics cleanser.

Chlorophyll: Any of numerous membrane-bound, porphyrin-based, photosynthetic pigments present in higher plants and algae.

Chromophore: Light-absorbing component of a pigment, such as the retinal in rhodopsin or the flavin in a flavoprotein.

Cilium: Short (5–10 micrometers) tail-like appendage of eukaryotic cells that is used for cell locomotion or feeding and consists of nine pairs of microtubules arranged around a central pair of microtubules.

Circadian rhythm: Biological rhythm that has a period of about twenty-four hours under constant environmental conditions, is generated endogenously but regulated exogenously, runs at approximately the same period at different temperatures, and can be reset by light and certain chemicals.

Compound eye: Eye that consists of many individual visual units and that occurs in insects and other arthropods (see figure 4, p. 36).

Cone cell: Cone-shaped, iodopsin-containing cell in the eye that is responsible for color vision in bright light.

Cyclic adenosine monophosphate (cAMP): Compound formed from ATP that has numerous biological roles, including intercellular communication in *Dictyostelium* and mediation of the effects of some animal hormones (see figure A10, Appendix).

Cyclobutane Pyrimidine Dimer (CPD): Most common type of ultraviolet-induced DNA damage, in which adjacent pyrimidine molecules are connected by a cyclobutane ring (see figure A3, Appendix).

Dinoflagellate: Aquatic (mostly marine), single-celled, flagellated, photosynthetic organism that has an intricately shaped outer shell, some species of which (e.g., *Gonyaulax, Gymnodinium*) are responsible for toxic red tides (see figure 10, p. 156).

Endosymbiosis: Union of two organisms in which one lives inside the other.

Energy transduction: Transformation of energy in one form into energy of a different form, such as photosynthesis, in which solar energy is transformed into biochemical energy.

Eubacteria: Major group of prokaryotes that are genetically and metabolically different from Archaea and include most gram-positive bacteria, cyanobacteria, mycoplasmas, enterobacteria, and pseudomonads (see figures 1, p. 4; and 11, p. 163).

Eukarya: Large group of single-celled and multicellular eukaryotic organisms, including plants, fungi, and animals, that have true nuclei containing genetic material (see figures 1, p. 4; and 11, p. 163).

Eukaryote: Cell that has a true nucleus, a distinct membrane-enclosed structure that contains genetic material.

Fermentation: Breakdown of energy-rich substances, such as

sugars, into simpler substances, such as ethanol and carbon dioxide, without the use of oxygen.

Flagellum: (1) Long (up to 150 micrometers) tail-like appendage of eukaryotic cells that is used in cell locomotion and feeding and, like a cilium, has nine pairs of microtubules arranged around a central pair of microtubules. (2) Tail-like appendage of prokaryotes that is composed of the protein flagellin and lacks the 9 + 2 arrangement of eukaryotic cilia and flagella.

Flavin (vitamin B-2): Any of several water-soluble, nitrogen-containing pigments that are derived from isoalloxazine, such as riboflavin, flavin mononucleotide (FMN), flavin adenine dinucleotide (FAD), and roseoflavin (see figure A9, Appendix).

Flavoprotein: Protein that has a flavin nucleotide covalently or noncovalently attached and is typically involved in oxidation/reduction reactions.

Ground state: Lowest energy state of an atom or molecule.

Halorhodopsin: Light-sensitive, retinal-based, membrane-bound protein that pumps chloride ions into *Halobacterium salinarum* cells (see figure 13, p. 170).

Heme: Compound that consists of an iron atom bound to a porphyrin ring, occurring in oxygen-transporting proteins, such as hemoglobin and myoglobin (see figure A5, Appendix).

Homology: Similarity of biological features due to inheritance from a shared ancestor.

Hydrothermal vent (black smoker): Chimneylike geological formation that emits heat and minerals from the earth's interior into the ocean.

Hypericin: Photosensitizing red pigment present in St. John's wort that is classified as a meso-naphthodianthrone-type molecule (see figure A11, Appendix).

Iso-alpha-acid: Compound derived from alpha acid during the wort boil, whose bitter taste balances the residual sweetness in fermented beer (see figures A7 and A8, Appendix).

Luciferase: Generic term for the enzyme that catalyzes the oxidation of luciferin and subsequent production of bioluminescence.

Luciferin: Generic term for the light-emitting compound of a bioluminescence reaction.

Mash: Mixture of milled grains and warm water in which enzymes break down starches to produce a solution of sugars (wort) that is subsequently fermented to make beer.

Mehler-peroxidase reaction: Set of reactions driven by superoxide dismutase and ascorbate peroxidase that removes certain toxic forms of oxygen (superoxide, O_2^-, and hydrogen peroxide, H_2O_2) and produces water as a harmless byproduct.

Melanoma: Tumor that arises from a melanocyte cell of the skin or eye and is the leading cause of death from skin cancers.

Melatonin: Hormone that is derived from serotonin and is secreted by the pineal gland of vertebrates, especially during darkness, and is involved in regulation of circadian rhythms (see figure A4, Appendix). Melatonin also occurs in *Gonyaulax* and some plants.

Mercaptan: Any of a group of chemicals (hydrosulphides of alcohol radicals) that are generally colorless liquids and have strong, disagreeable odors (see figure A7, Appendix).

Morphogenesis: Development of biological form, including the formation of tissues and organs.

Myoglobin: Red, heme-containing, oxygen-transporting protein that is present in the muscles of vertebrates and some invertebrates.

Norflurazon: Herbicide that blocks synthesis of carotenoids.

Northern hybridization: Molecular biology technique used to determine the level of a specific RNA molecule in a biological sample by measuring its hybridization (binding) to a radioactively labeled strand of similar RNA or DNA.

Ommatidium: One of many optical units that constitute the compound eye of such arthropods as mantis shrimps, butterflies, and shrimp (see figure 4, p. 36).

Opsin: Protein component of visual pigments in the rhodopsin family that is typically associated with retinal or a related compound.

Phosphodiesterase: Enzyme that breaks down cyclic AMP to AMP (see figure A10, Appendix).

Photobiology: Branch of biology that deals with the biological effects of light, from the ultraviolet to the near-infrared.

Photodermatitis: Inflammation of the skin caused by exposure to light.

Photo-immunosuppression: Suppression of the immune system by light, especially ultraviolet radiation.

Photo-isomerization: Light-induced change in the structure of a chemical compound without a change in the number of atoms, such as occurs in phytochrome (see figure A6, Appendix) and rhodopsin (see figures A1 and A2, Appendix).

Photokinesis: Light-induced change in cell speed or frequency of alterations in cell direction that results from sensation of the photon fluence rate ("intensity") of light.

Photolyase: Enzyme (Cyclobutane pyrimidine dimer photolyase or pyrimidine [6–4] pyrimidone photolyase) that repairs damaged sites in DNA when activated by light.

Photolysis: Breaking of chemical bonds by light or electromagnetic radiation (see figure A7, Appendix).

Photon: Indivisible unit (quantum) of light or electromagnetic radiation.

Photon flux: Number of photons that strike a flat surface of unit area per unit of time, e.g., 100 billion photons per square meter per second ($10^{11} \, \mathrm{m^{-2} \, s^{-1}}$).

Photooxidative stress: Biological stress that results from the light-driven generation of toxic forms of oxygen, such as singlet oxygen ($^1O_2$).

Photophobic response: Photon fluence rate ("intensity") change–induced alteration of cell movement, typically manifested as a stop and reversal of direction.

Photosensitization: Process in which absorption of radiation by one molecule leads to chemical alteration in one or more other molecules.

Photosynthesis: Biological conversion of light energy into biochemical energy that occurs in most plants and algae, where it is mediated by chlorophyll, and in some prokaryotes, where it is mediated by other pigments, such as bacteriochlorophyll or bacteriorhodopsin.

Phototaxis: Movement by a motile organism with respect to the direction of light that is typically toward (positive), away from (negative), or perpendicular to (transverse) the light source (see figure 8, p. 130).

Phototropism: Light-directed orientation of a sessile organism or part of an organism that involves turning or curvature by differential growth (see figure 6, p. 107).

Phytochrome: Photoreceptive plant pigment that regulates many stages of growth and development, including seed germination, stem growth, flowering, and senescence (see figure A6, Appendix).

Pineal gland: Small pinecone-shaped appendage of the brain of vertebrates that is involved in regulation of circadian rhythm; photosensitive in certain fishes, amphibians, and reptiles.

Polarization: State of radiation in which the oscillation of the waves, when viewed in the direction of light propagation, have a pattern that is classified as linear (if the endpoint of the electric vector moves along a straight line), circular (if the endpoint of the electric vector moves along a circle), or elliptical (if the endpoint of the electric vector moves along an ellipse).

Porphyria: Any of several hereditary or chemically induced abnormalities of porphyrin metabolism, some of which are characterized by excretion of excess porphyrins or porphyrinogens in the urine or feces, neurological disorders, and sensitivity to light (see table A1, Appendix).

Porphyrin: Any of numerous cyclic compounds, such as chlorophyll or heme, that has four pyrole rings joined by four $=CH-$ groups and a central metal atom (see figure A5 and table A1, Appendix).

Porphyrinogen: Any of numerous porphyrin-like compounds in which the pyrole rings remain in their reduced state (see figure A5 and table A1, Appendix).

Prokaryote: Cell that lacks a membrane-enclosed nucleus and is classified as an Archaea or Eubacteria.

Retinal: Aldehyde of retinol that is derived from dietary carotene and binds to the protein opsin to form visual pigments in the rods (rhodopsin) and cones (iodopsin) of the eyes of most animals (see figures A1 and A2, Appendix).

Retrovirus: Virus such as HIV (human immunodeficiency virus) that has RNA as its nucleic acid and uses reverse transcriptase to copy its genome into the DNA of host cells.

Rhodopsin: Photoreceptive pigment of animals that consists of retinal bound to opsin and is used for vision in dim light.

Rod cell: Rod-shaped, rhodopsin-containing cell in the eye that is responsible for vision in dim light.

Saccadic movement: Rapid eye movement that allows redirection of a gaze without blurring the visual image.

Scintillon: Special organelle (membrane-enclosed structure within a cell) present in *Gonyaulax polyedra* and some other dinoflagellates that contains all the components necessary for bioluminescence.

Sensory rhodopsin: One of two retinal-based photoreceptive pigments (sensory rhodopsin I or II) in *Halobacterium salinarum* and some other Archaeans that controls swimming away from ultraviolet and blue light and toward orange-red light.

Serotonin (5-hydroxytryptamine): Neurotransmitter and hormone that is synthesized from tryptophan; involved in diverse physiological processes such as regulation of mood, circadian rhythm, and blood vessel constriction in vertebrates, invertebrates, plants, and other organisms (see figure A4, Appendix).

Sexual selection: Evolutionary process that depends upon the reproductive advantage that some individuals have over others
of the same gender and species, and results in the enhance-
ment of traits involved in mate acquisition.

Singlet state: Chemical state, typically short-lived, in which the
total electron spin quantum number is zero.

Slug: Sausage-shaped multicellular structure of cellular slime molds,
also known as the pseudoplasmodium (see figure 7, p. 118).

Sorocarp: Spore-bearing structure of cellular slime molds (see
figure 7, p. 118).

Southern hybridization: Molecular biology technique used to
search for a specific segment of DNA in a biological sample
according to its hybridization (binding) to a radioactively
labeled segment of similar DNA.

Spherical photon flux: Number of photons that strike a sphere of
unit cross-section per unit of time, e.g., 100 billion photons per
square meter per second (10^{11} m^{-2} s^{-1}; see table 1, p. 6).

Sporangiophore: Stalk that bears the reproductive spores of *Phycomyces* and other Zygomycete fungi (see figures 5, p. 106; and
6, p. 107).

Squamous cell carcinoma: Tumor that originates from a squamous cell on the skin, lips, mouth, throat, or esophagus.

Stentorin: Hypericin-like pigment that controls cellular movement of *Stentor coeruleus* (see figure A11, Appendix).

Stratosphere: Upper region of the atmosphere that contains most
of the atmosphere's ozone (O_3).

Suprachiasmatic nuclei (SCN): Region of the hypothalamus in
the brain that lies just above the optic chiasm, where certain
neural fibers from the left and right eyes cross one another.

Threshold: Level of stimulation by light or another factor that
causes a barely detectable response.

Transduction: Conversion of a signal from one form to another,
such as the conversion of light into nerve impulses by the eyes
and central nervous system.

Triplet state: Chemical state, typically short-lived, in which the total electron spin quantum number is one.

Vision: Sense mediated by the eye and central nervous system that allows perception of the appearance of objects.

Wavelength: Distance between two corresponding points on adjacent waves when measured from the direction of propagation, which, for visible radiation, determines the color (see figure 2, p. 9).

Wort: Solution rich in sugars that is formed by removal of the spent grains following completion of the mash and is fermented to make beer.

Xanthophyll: Any of several red, orange, or yellow carotenoid pigments that is an oxygenated derivative of a carotene (see Figure A14, Appendix).

Index